GOVERNMENT AND INNOVATION

PETRA SÖDERLING

GOVERNMENT AND INNOVATION

THE ECONOMIC DEVELOPER'S GUIDE TO OUR FUTURE

Government and Innovation

- The Economic Developer's Guide to Our Future

Petra Söderling

Petra Söderling/books
http://petrasoderling.com
petra@petrasoderling.com

Library of Congress Cataloging-in-Publication Data:
Söderling , Petra
Government and Innovation - The Economic Developer's Guide to Our Future

ISBN 979-8-9881124-0-2 (Ebook)
ISBN 979-8-9881124-2-6 (Hardback)
ISBN 979-8-9881124-1-9 (Paperback)

2023

Cover and layout design by Marko Markovic, 5mediadesign.com.
Editor Tamma G. Ford.

Manufactured in the United States.

CONTENTS

SECTION 1

TOOLS

CONTENTS

PREFACE

Dr. Martin E.P. Seligman is an American psychologist who has written multiple books about hope and optimism. His argument is, by shifting the focus from pessimism and negative thoughts consciously towards positive thinking, psychology and psychologists can achieve better results in treating and embracing mental health.

Politicians, civil servants, voters, tax payers, we're all human. We're not immune to Seligman's theory. I'm writing this book to give hope and optimism for cities, states, countries and alliances on how investing in innovation can shift economic development towards a chosen goal. This may sound like a no brainer, but the trick is how to approach it so it works for your constituency. If you are currently in public office as a civil servant or elected representative, running, or considering running for office, or otherwise creating strategies on public policy, this book is for you. If you are a voter or a taxpayer, this book is for you, too.

In my roles within a multinational technology corporation and a startup founder and CEO, I had never given much thought to the role of public policy. I was too busy building the next big thing, a gadget or a solution that would revolutionize how people do some mundane task, like pay their bills, play games, post pictures online, or buy a house. These were, in my younger mind, important innovations that would radically take humankind to the next level. All we needed was investors and engineers, we knew what to do and didn't need anyone to tell us otherwise.

The closest to the government's role was in the early 2000s when my team was eagerly waiting for European governments to auction 3G spectrum licenses so we could start selling 3G networks and devices. As a fast-growing exciting new tech industry, we took it for granted that the governments are doing this because we - the industry - demanded it from them. We knew what we were doing after all and didn't want to hear any counter arguments.

It wasn't until I was hired by one of these European governments that I started to pay attention to its role and impact in technologies. Could governments actually be strategic, active players in creating new industries? Could it be that had it not been for prior art paid by the government, none of us private companies would have been able to develop the 3G standard.

As I continued my career into the public sector, I came across problems that seem hard for many to solve. And they are hard. If it would be easy for a country to invest in innovation, then just stand by and witness economic growth, everyone would be doing it - successfully. There would be no poverty, no underperformers. In addition to asking why, or why not, many are asking how. How do I turn my poorly performing country, state, or city into a high performing one? I don't have any extra money in my budget, and I can't just raise taxes. Should I take a loan? Can I get a loan? Maybe I need a boost from someone not living here, a Foreign Direct Investment (FDI). But who would invest in my region and why?

These are real questions posed every day by real people in governments all over the world. Compared to how many books are written for business leaders guiding them through tough decisions, there are way too few books on helping public servants make the most of their economies. This book is written as an easy read, hopefully as enjoyable as informative, so it could find its way into the hands of as many people as possible.

Public decision making has influenced, and continues to influence, the industries and commercial breakthrough successes we have around the world. There are people behind those decisions, some people make some of these decisions public, some decisions are made behind the scenes by people you've never heard of. This book will look at some of the most successful countries, including Canada, Brazil, Finland, Italy, and South Korea, how their innovation funding policy has shaped that success, and who were the people behind those decisions.

Often in life, a big shift, improvement, or disruption can be traced back to one person. This holds true for human rights achievements, environmental laws, a technical gadget, scientific breakthrough, as well as for a new bridge or a baseball field. I want you to be that person, and I want you to approach this problem with hope and optimism.

Dr. Seligman's famous theory of well-being answers the question "what is human flourishing and what enables it "[1]? He lists five building blocks (PERMA; Seligman, 2012): Positive Emotion, Engagement, Relationships, Meaning and Accomplishment. Here's a good description I borrowed from University of Pennsylvania, Penn Arts & Sciences workshop on PERMA™ theory of well-being:

> *"Positive Emotion: This route to well-being is hedonic – increasing positive emotion. Within limits, we can increase our positive emotion about the past (e.g., by cultivating gratitude and forgiveness), our positive emotion about the present (e.g., by savoring physical pleasures and mindfulness) and our positive emotion about the future (e.g., by building hope and optimism).*
>
> *Unlike the other routes to well-being described below, this route is limited by how much an individual can experience positive emotions. In other words, positive affectivity is partly heritable and our emotions tend to fluctuate within a range. Many people are, by disposition, low in experiencing positive emotion. Traditional conceptions of happiness tend to focus on positive emotion, so it can be liberating to know that there are other routes to well-being, described below.*
>
> *Engagement: Engagement is an experience in which someone fully deploys their skills, strengths, and attention for a challenging task. According to Mihaly Csikszentmihalyi, this produces an experience called "flow" that is so gratifying that people are willing*

1 PERMA™ Theory of wellbeing, Positive Psychology Center, Penn Arts and Sciences

to do it for its own sake, rather than for what they will get out of it. The activity is its own reward. Flow is experienced when one's skills are just sufficient for a challenging activity, in the pursuit of a clear goal, with immediate feedback on progress toward the goal. In such an activity, concentration is fully absorbed in the moment, self-awareness disappears, and the perception of time is distorted in retrospect, e.g., time stops. Flow can be experienced in a wide variety of activities, e.g., a good conversation, a work task, playing a musical instrument, reading a book, writing, building furniture, fixing a bike, gardening, sports training or performance, to name just a few.

Relationships: Relationships are fundamental to well-being. The experiences that contribute to well-being are often amplified through our relationships, for example, great joy, meaning, laughter, a feeling of belonging, and pride in accomplishment. Connections to others can give life purpose and meaning. Support from and connection with others is one of the best antidotes to "the downs" of life and a reliable way to feel up. Research shows that doing acts of kindness for others produces an increase in well-being.

From an evolutionary perspective, we are social beings because the drive to connect with and serve others promotes our survival. Developing strong relationships is central to adaptation and is enabled by our capacity for love, compassion, kindness, empathy, teamwork, cooperation, self-sacrifice, etc.

Meaning: A sense of meaning and purpose can be derived from belonging to and serving something bigger than the self. There are various societal institutions that enable a sense of meaning, such as religion, family, science, politics, work organizations, justice, the community, social causes (e.g., being green), among others.

Accomplishment: People pursue achievement, competence, success, and mastery for its own sake, in a variety of domains, including the workplace, sports, games, hobbies, etc. People pursue accomplishment even when it does not necessarily lead to positive emotion, meaning, or relationships."

I'm not using the PERMA model as a framework in this book, but the idea of learned optimism, shared goals, and meaningful accomplishments are certainly valuable tools under our belts. Increasing positive emotion, seeking the flow, and connecting with others in order to serve something bigger than yourself has been the enablers for me to get this book written and published.

In August 2021 I had just finished reading Mariana Mazzucato's Entrepreneurial State, and I was excited about her message on governments being active industrial spearheads. To support the conversation I launched a podcast called Deep Pockets, available on Apple, Google, and Spotify podcasts. At first my goal was to uncover as many technologies as possible as being publicly funded. I soon realized this was a naive goal, as *all* technologies were originally in one way or another publicly funded. I then started adding episodes on economic development projects, how places such as New Orleans, LA could reinvent themselves with the use of innovation. This, too, started feeling like it's not adding enough value. As I thought of this more and more, I decided a podcast is not a good vehicle for sharing complicated thought processes on money, policy, and technologies. I needed to put these ideas on paper, and here we are.

Many of the interviews in this book are from those podcasts, and many more podcast episodes ideas will emerge from this book. I hope you enjoy reading this book, and listening to the Deep Pockets podcast. My social media channels are always open for constructive feedback, new ideas, or project proposals.

FOREWORD

by Rob Hays, CEO of Atom Computing

Government and innovation are two words not often spoken back-to-back and yet they are intimately linked to each other and the prosperity of societies as Petra Soderling so eloquently illustrates for us. "Government and Innovation" is like a gamer's guide to understanding the role government, at all levels, plays in creating successful, innovative economies where people can thrive. By providing historical context and modern examples, Petra provides a tangible set of tools that policy makers and business leaders can use to spur innovation in their geographies. These are lessons that I've learned on the job "the hard way" as I continue to uncover how intertwined government and innovation really are.

As a product of public schools and university, I understood the role of government in providing education and scientific research. However, as I began my career, I had no idea how vast the role of local and national government forces were in influencing where, what, and how innovation takes shape. I used to think that companies were on their own to develop products and compete in the global market with the only real government intervention showing up in earnings releases in the form or taxes. I believed the global market was open, fair, and largely free of government intervention. "The world is flat", after all.

The day after I graduated from Georgia Tech with a Computer Engineering degree, I hopped in my car and headed west on my first transcontinental road trip. Like the 49ers a century and a half before, I left my native state behind and steered toward California for new opportunity. I had an exciting new position lined up at one of Silicon Valley's iconic companies, Intel. Having studied microprocessor design, I decided that joining the world's leading microprocessor company was a reasonable place to start. Wisely, I left a month of free time between school and work. Before diving headfirst into my career, I wanted to explore some of the western National Parks – one of the world's greatest *government innovations*.

After crawling through Carlsbad Caverns, rafting the Grand Canyon, and hiking the Sierra Nevada on my way to my new home in California, I settled in as an engineer at Intel. It was truly a great place to work. We were not just riding the wave of Moore's Law, we were inventing it. Our co-founder was Gordon Moore himself. At that time, our team produced 100 million processors annually[1], refreshed the design every year, and developed a new manufacturing process every other year. I was part of an innovation machine that was at the foundation of a global ecosystem of innovation in systems, software, and services. Intel processed silicon wafers primarily in U.S. fabs and sent them to the Philippines and Malaysia for packaging and final test. Motherboards and chassis were almost exclusively produced by multiple vendors in Taiwan. The memory and hard drives mostly came from Japan and South Korea. OEMs around the world assembled the computer systems and marketed them under their own brands. Iconic companies were built on top of Intel's x86 microprocessor platform. While we were cranking out processors and enabling generational paradigm shifts like the Internet, mobility, and cloud computing, I was blissfully ignorant of the supporting role that government played.

Over the course of my career at Intel, I moved to different business units and locations around the country, eventually becoming a Vice President leading the product roadmap and innovation for the data center business unit in Oregon. As an executive, I became much more aware of geopolitical influence, government research, tax incentives, safety and environmental regulations, international standards, and other ways that government actions help shape business decisions and the growth, or absence, of economic development in various regions. We're not talking about small amounts of support.

The city of Hillsboro, Oregon and Washington County approved a 30-year deal with Intel granting up to $2 billion in property tax breaks on up to $100 billion in local investment in semiconductor fabs[2]. Similar tax incentives were granted to Intel to update their fabs in Arizona, Ireland, and Israel. Intel's sizable moat was continuously deepened with the help of local tax incentives to offset the massive cost of the new fabs required to fuel Moore's Law and the tech innovation that depends on it across all industry sectors. Of course, that value was returned to the local economies in the form of jobs, income tax, property tax, sales tax, and spending. As the largest private employer in the state of Oregon, Intel's positive impact on the economy cannot be underestimated.

At the same time, a looming challenge to global semiconductor leadership emerged from the east. In 2014, China released their 12th 5 year plan declaring domestic semiconductor design and manufacturing as national priority[3]. China's industrial and societal dependency on foreign semiconductors was pervasive. Semiconductors had grown to rival oil as the top import in China[4], with the top three semiconductor manufacturers being Intel, TSMC, and Samsung. Many of the world's electronics assembly factories had migrated from Taiwan to China over the previous decade making a significant portion of China's industrial economy dependent on the foreign semiconductor supply chain. Additionally, Chinese technology brands had grown into global leaders – Huawei, Lenovo, Xiaomi, Ten Cent's WeChat, Alibaba, and Byte Dance's TikTok. All of these companies had transformed the way people in China and around the world communicated, socialized, shopped, and worked. All of them were also completely dependent on foreign semiconductors. China's aim was to address the dependency by growing a competitive, domestic semiconductor capability by 2025.

China's declaration was succeeded by significant investments in their domestic semiconductor R&D infrastructure including notably academic research with in their technical universities, an infusion of capital into their #1 foundry, SMIC, and a wave of joint ventures between Chinese state-owned entities and international semiconductor companies. These joint ventures promised the opportunity of access to the Chinese market in exchange for a share of the profits and, more importantly to China's long-term aspirations, license to intellectual property that could help jump start the long road to semiconductor independence.

For several years, I experienced firsthand frustration as the U.S. government left companies alone to figure out how to compete in China under increasing pressure. As you would imagine, companies that had a deep cache of trade secrets and strong market share were in a better position to negotiate a "free market" in China than companies in weaker positions. Those with small share in China were more eager to form joint ventures as an upside growth opportunity because they had nothing to lose. We were in a trade war that I don't believe was well understood publicly. It was obvious to those of us on the inside that help from our federal government was required to stem the outflow of critical IP and tip the balance of trade vs. China's aspirations. Ten years after the Chinese declaration, the U.S. CHIPS and Science Act of 2022 declared the strategic importance of semiconductor leadership and committed over $50 billion in capital investment in R&D and manufacturing projects in America[5].

After over twenty years at Intel, visiting customers around the world, it had become clear that something different was happening in China. Not just the semiconductor trade war, but more interesting to me, Chinese innovation was transforming their society at rate that could be observed over a few years not just decades. Walk into any coffee shop or store a decade ago and everyone paid with their phones using WeChat Pay or Alipay. Apple Pay, a similar concept, is just starting to penetrate retail in the U.S. The motherboard, PC, and phone factories in China are enormous – literally city size – and employ tens of thousands of workers who live in gigantic cities that seemingly spring up overnight.

One of the first and most recognizable global brands to emerge out of China was Lenovo – one of the largest computer companies and, as so naturally, a top customer of Intel. When they recruited me to be Chief Strategy Officer of their datacenter business unit, I was intrigued to join a company with Chinese origin and a global growth business. This position afforded me an entirely new and more well-rounded perspective of how international government roles influence business and innovation decisions. Lenovo's growth from a start-up PC manufacturer in China into one of the world's leading technology providers from smart phones to supercomputers, was a result of decades of internal innovations in design, manufacturing, supply chain, and marketing plus mergers and acquisitions of rival businesses like IBM's PCs and x86 server businesses and Motorola phones from Google.

However, continuing growth through acquisition was dampened by the Committee on Foreign Investment in the United States (CFIUS). CFIUS is responsible for reviewing major investments and change of control of U.S. companies by foreign entities. By increasing the scrutiny of investments by Chinese companies, the time and cost required to complete the review process had effectively dissuaded many companies from pursuing them.

At the same time, tariffs and export controls sparked by growing nationalism in the U.S. were heating up the trade war and driving a "made in China, for China" sentiment among the Chinese government and citizens. The combination of CFIUS and import/export sentiment affected our leadership team's thinking on what options we should consider for our growth strategy and influenced us to turn more heavily on internal innovation – make vs. buy – to continue our growth. Much of that innovation would occur in China, for China, while we would continue to design and build our global products in other counties including the U.S., Japan, Taiwan, Mexico, and Hungary among others.

Now as CEO of Atom Computing, a company building quantum computers, I am peeling new layers of the onion in learning how government influences innovation. This time it's less about figuring out how to navigate regulations, export controls, taxes, and tariffs and more about how to engage the government directly in research collaborations and sponsorships to accelerate our innovations in support of their mission. Quantum computing has the potential to transform many industries with unprecedented computational capability to simulate complex physical systems, but it will take many years of technical advancements to mature today's small-scale systems into fault-tolerant systems with large-scale commercial value.

The U.S. government is one of our primary sponsors through grant programs designed to help our company and others develop the technologies required to achieve our objectives. We're also collaborating with an ecosystem of vendors in Europe and Asia to advance multiple layers of the solution stack from components up to algorithms and applications. Many of our partners are also sponsored by their local governments to support their efforts to establish technical leadership in their jurisdictions for economic development and national security purposes. I'm discovering how beneficial public-private partnerships are in tackling difficult, long-term technical challenges in pursuit of a common goal.

It was in this role that I met Petra Soderling – a woman who wears many hats. Her position at the World Bank has given her unique and broad purview over the role that governments and fiscal policy play in shaping economic leadership. She has taken a personal interest in quantum technologies and works with companies like ours to help advise our strategy on government collaborations. She was also host to a delegation of Colorado companies, that Atom Computing was part of, as we visited Finland to explore business collaboration opportunities there.

While it is a big part of my experience, this isn't a book about U.S.-China relations. This is a book about how policy makers can create an environment of innovation that will help their economies and citizens prosper. It is also a book to help business leaders decipher the complicated world of policy and take better advantage of the government support available or help shape it. Petra uses real-world examples from different continents and industries to illustrate how a strategic focus on an area of future promise, finding ways to encourage broad participation, and providing adequate funding can spawn an ecosystem of innovation. As a reader, I appreciate both the specific examples and the concise take aways summarized after each section.

One thing that has become certain upon reflection of my journey and reading Petra's book is that the role of government is vast and pervasive in spurring innovation and fostering an environment for it to flourish. Understanding the game and its players takes considerable time and effort. It can be learned over the course of a career and even then, there are still mysteries and unknown forces at play. Your learning can be accelerated by reading this book and applying this practical advice to promote government and innovation in your area. Adam Smith's "invisible hand" that that helps autoregulate pricing due to laws of supply and demand is referenced in the book. Similarly, Petra's insights help the reader understand how the invisible hand of government can play an active role in shaping a culture of innovation and prosperity.

SOURCES:

Nona Yates, Dell Passes Compaq as Top PC Seller in U.S., Los Angeles Times, January 24, 2000

Luke Hammill, Intel's tax breaks: Historic 30-year deal approved by Hillsboro, Washington County, Oregon Live, August 27, 2014

John VerWey, Chinese Semiconductor Industrial Policy: Past and Present, United States International Trade Commission Journal of International Commerce and Economics, July 2019

Electronic circuits; monolithic, integrated, digital, metal oxide semiconductors (MOS technology), The Observatory of Economic Complexity (OEC)

CHIPS and Science Act Will Lower Costs, Create Jobs, Strengthen Supply Chains, and Counter China, The White House, August 9, 2022

INTRODUCTION TO SECTION 1

The latter part of the 2010s and first years of the 2020s have seen an increasing amount of political divide, partisanship, and the rise of extreme views. In the US and Europe, right wing movements gained traction from problems arising out of immigration. In Europe, it was first fueled by asylum seekers from the war in Syria, and in the US by immigrants arriving from Central and South America. Radical right, neo-Nazi parties emerged in the Nordic countries, Austria, France and Germany, spreading hate online, while their more toned-down sister organizations were elected to parliaments across Europe. Environmental groups, especially in Europe, adopted the playbook of Greenpeace and started to block highways and main streets, harming private and public assets. American paramilitary groups turned their focus from defending the nation against possible foreign adversaries to organizing around domestic issues, becoming increasingly violent and dangerous.

Social media companies realized they can sell more ads with divisive content and started to favor content that provides extremist viewpoints. When Donald Trump won the presidency despite having demonstrated misogyny, xenophobia, and racism during his campaign, many felt that it was now acceptable to break with the prevailing norms and say out loud what could not have been said before.

This newly felt "freedom"—freedom from norms and in some cases freedom from truth—spread also to views on how to run economies. When governments had to close parts of society due to what was an unknown, highly transmittable and deadly virus, millions of people felt that things had gone too far. Democratic governments had overstepped their boundaries. Instead of using the power that people gave them to govern, they had turned into tyrannies and dictatorships. Not only did people not want to take a vaccine that was quickly developed by publicly subsidized giant companies, but they did also not see why the government would need to order private restaurants and gyms to close. What right do they have?! Wouldn't it just be better for everyone if we allowed businesspeople to run their businesses as they see fit, and have those bureaucrats step out of the way once and for all?

Well, no, actually.

I will be laying out a number of good reasons why we need government, and how the best economies are actually the ones with a healthy democracy and a strong government. They are the ones where economies are driven by innovation, and where this innovation is a coordinated result of private and public organizations alike.

Countries or states that have a vision and a strategy and an associated action plan for where they are headed, what kind of a society they want for their grandchildren, and also know, what tools and resources are needed in order to get to those goals, are on top of global indexes that measure prosperity, resilience, well-being, and happiness. This vision, strategy, and action plan (and the tools) can only be created by a public body. Private companies are responsible for creating visions, strategies, and action plans for their owners and other stakeholders. It is not a good idea to allow a company to start shaping the future of an entire country, because that could lead those voters and taxpayers to become nothing but free resources for the shareholders of one private enterprise. Much better idea to keep voting for people who can articulate a vision, and work together with other democratically elected officials to work towards that vision.

In the first section of this book, we will discuss a number of tools that national, regional, and local governments have at their disposal. The list is not exhaustive, of course, but it is selected to demonstrate how many ways public officials possess to influence innovation-based prosperity in their region. Some are more long-term than others, but that is the beauty of government-led, or government-influenced innovation; it can invest a lot more resources and stay the course during a lot longer turnaround than any private investor could.

I've already used the term 'prosperity' a few times, and it's perhaps a good moment to define what I mean by that. I am arguing that by investing in innovation enablers, governments can generate prosperity for their voters and taxpayers, residents and citizens. Functioning markets, GDP, corporate profits and shareholder value can all lead to prosperity, but they are– I argue – neither a measure of prosperity nor prosperity itself. Prosperity in this context is wealth that works for the entire economy and is felt and enjoyed if not all, then almost all of the people living in that society. Wealth, in our vocabulary, includes elements like salaries and assets, but also health and happiness in people and nature. Prosperity generates more prosperity; it is never a dead end.

These tools – prosperity enablers if you will - are also not surgical tools that you can introduce and take away without short- or long-term consequences. Many have a federal or international impact and are dependent on how others react to you using this tool. We all live simultaneously in our local economies and the global economy. As an example, the USA recently introduced massive government subsidies for American families to buy electric vehicles that are made in the USA. This sounds like a good initiative for American families and American car makers. However, car makers in other countries saw it as a hostile move towards them, which may result in their local governments to issue a similar package in their regions. This could circle back to hurt American car manufacturers who would like to sell their cars in Europe and Asia.

The tools are also interdependent. Good daycare and education system attracts young families to move into your area, the more children and young adults graduate from good schools, the more companies want to set up operations there, the more companies and jobs, the more adjacent services that make your town a fun place to be, the more tax dollars to pay for roads, fast internet, public parks, events, libraries and research. You can take one of the building blocks away and things will run smoothly for a while, but at some point the machine will start coughing and your upward cycle will stall. Families start moving away for better opportunities, and your town will become a shadow of its glorious past. We've all seen those places.

So no, we cannot replace the government with an app, like someone has suggested. We also cannot replace the government with Artificial Intelligence, or any other technological gadget. Government acts on behalf of the people for the common good of the people, sustainably long term.

SECTION 1

TOOLS

HOW GOVERNMENTS DRIVE INNOVATION

Historically, economic success in western democracies has been attributed to the free market. This economic system famously coined by Scottish philosopher Adam Smith in his 1776 book 'Wealth of Nations' was the cornerstone in building the United States of America, also born in 1776.

Adam Smith brought forward the idea of a market where anyone can buy and sell what they like at a price that someone is willing to pay. Here, the laws of supply and demand autoregulate pricing, which is what he called the 'invisible hand'. Adam Smith favored assembly-line production where people are divided into tasks for maximum efficiency. He is also credited with inventing the gross domestic product (GDP) as a great tool for measuring output and this newly discovered efficiency. GDP, as old as it is, is still considered to be the most comprehensive measure of economic activity. GDP measures the value of the final goods and services produced. Changes in GDP are the most popular indicator of a nation's overall economic health. Lately GDP has received criticism because it does not consider the wellbeing of citizens, pollution, or diminishing natural resources.

Another big name in free market economics is Milton Friedman. He was born in 1912 in New York City as the fourth child to Jewish immigrants Sarah Ethel and Jeno Saul Friedman, who had escaped hardships in Austria-Hungary. The family income was, in Friedman's own words, "small and highly uncertain; financial crisis was a constant companion"[2]. Young Milton attended public elementary and secondary school and was able to enroll into university with a competitive scholarship from the State of New Jersey. He was to play an important role in America becoming the 20th century miracle that it is.

As an economist, Milton Friedman popularized the term 'monetarism', the quantity theory of money. According to monetarism[3], the best way for governments to affect GDP is by controlling the amount of money circulating in the economy. Governments do this by adjusting interest rates (see more in the chapter on Interest Rates and Taxes). Critics say monetarism rejects other equally good instruments, many of which we are laying out in this book. Friedman also came up with the idea of 'shareholder value', a school of thought where the sole purpose[4] of a firm is to make money for its shareholders. This idea, too, was later criticized.

Milton Friedman also served in many US administrations, including that of Ronald Reagan, who famously popularized the idea of 'trickle down economics'. This is the idea that as rich get richer, their wealth somehow trickles down to poor people—also a highly criticized if not outright debunked idea. Friedman's theories and teachings have had a profound and long lasting impact on the American economy, and also affected economic policy makers elsewhere, perhaps most famously in the United Kingdom (UK) under Margaret Thatcher.

Over the course of the 20th century many European countries, not just Thatcher's UK, looked at the US and thought to themselves, "That seems to be working, we want what they're having." The situation in Europe was different because of the setbacks suffered in two world wars, of course, so none of the European countries were ever able to create an economy quite like the US.

If you've noticed, I'm using the terms 'economy', 'market', and 'free market'. I am intentionally not using the term 'free market *capitalism*'.

2 Biographical of Milton Friedman, The Nobel Foundation 2005

3 Sarwat Jahan and Chris Papageorgiou, What Is Monetarism?, the International Monetary Fund, March 2014, Vol. 51, No. 1

4 Steve Denning, The Origin Of 'The World's Dumbest Idea': Milton Friedman, Forbes

This is not a book to criticize capitalism; I love capitalism. Growing up in a country neighboring the Soviet Union I saw first-hand how the other system, socialism, works. No thank you. This book is about what leaders can do to make their countries more successful through innovation, and also about proving that the so-called "free market" will not get you there. The free market has been myth busted many times before, but I'll just do it once more just for fun: There really is no such thing. Today, in 2023 and not 1776, Encyclopædia Britannica defines free market as a "market with no government intervention, no state rules and regulations, no taxes, no unions"[5]. This is what I would call a failed state.

The American non-profit, non-governmental research and educational institution The Fund for Peace, founded in 1957, works to prevent violent conflict and promote sustainable security. They publish a Fragile State Index[6], measuring fragility in 179 Countries. The most recent list of failed states[7] gives you these top ten "free markets": Yemen, Somalia, South Sudan, Syria, Democratic Republic of Congo, Central African Republic, Chad, Sudan, Afghanistan, and Zimbabwe. On the other end of the spectrum you will find countries like Switzerland, Finland, Norway, Denmark, Canada, and Sweden—all heavily taxed and regulated, and on top of the world in terms of their ability to provide for their citizens and protect their borders. These countries are the ultimate valedictorians in the global economy class.

But are they innovative?

Merriam-Webster defines innovation as "a new idea, method, or device."[8] How do you then measure 'new idea, method, or device'? Consultants are making lots of money presenting innovation measurement tools for corporations, some of them more accurate and useful than others. These include things like employees' perception of their employers' innovativeness, the amount of time employees "innovate" versus doing their regular job, revenue share from innovative products versus regular products, and so on.

5 Marc Orlitzky, free market, Britannica

6 Sarwat Jahan and Chris Papageorgiou, Back to Basics: What Is Monetarism?, Finance & Development, International Monetary Fund, March 2014

7 Failed States 2022, World Population Review

8 Innovation, Merriam-Webster

For a country or a state, things could get a bit blurry. In order to put investments into good use, the international community has agreed on a handful of metrics that are considered useful.

A metric that has grown in popularity is the 'percentage of public and private research and development (R&D) money of the overall GDP or a state budget.' The United Nations Educational, Scientific and Cultural Organization (UNESCO)[9] tracks these numbers for countries and presents them through different filters. One such filter is the number of people in that country working in R&D. UNESCO even offers this information categorized by gender.

The World Intellectual Property Organization (WIPO) lists on their website patents, trademarks and industrial designs filed by different countries.[10] You can do searches based on file types, industries, geographies, and technologies. WIPO's annual Global Innovation Index also takes into account the number of scientific papers published and the amount of risk investment money available within the country.

It's probably not a surprise when I say that the same countries that are highly regulated, have high tax rates, and possess a unionized workforce also top the statistics in most R&D spending, most people working in R&D, and most patents and other intellectual property filed and approved.

In other words, high regulation correlates with high innovation.

But is this merely correlation, or could there be causation involved? Italian-American economist Mariana Mazzucato argues for causation. In her book 'The Entrepreneurial State' she lays out a direct path from government investment to a successful, innovative private sector. Without the US government we would not have Google, SpaceX, or Pfizer. Dr. Mazzucato thinks taxpayers should be treated as risk investors, since many of the most successful products rely entirely on groundbreaking science that was publicly funded. I tend to agree. These types of investments—the internet, a space program, a biomedical science program—are

9 Welcome to UIS.Stat, The UNESCO Institute for Statistics (UIS)

10 Global Innovation Index 2021: Innovation Investments Resilient Despite COVID-19 Pandemic; Switzerland, Sweden, U.S., U.K. and the Republic of Korea Lead Ranking; China Edges Closer to Top 10, The World Intellectual Property Organization (WIPO), September 20, 2021

Top ten innovative countries 2021 by WIPO.

1. Switzerland
2. Sweden
3. United States of America
4. United Kingdom
5. Republic of Korea
6. Netherlands
7. Finland
8. Singapore
9. Denmark
10. Germany

Image credit Petra Soderling.

Top Ten innovative countries 2021, World Intellectual Property Organization (WIPO)

so expensive and require such a long lead time they could not possibly be done by the private sector.

When a Venture Capital (VC) firm invests in startups they take a big risk. Typically more than half of the startups they invest in will fail to some degree. Some fail completely, and some become stars of the portfolio by raking in more profits than was lost with the failing investments combined. This is why a VC fund wants to diversify its portfolio; it invests in different types of companies in order to maximize the possible outcome.

If you compare this with what countries are doing, it's not that different.

The most innovative economy in 2021 as measured by WIPO is Switzerland. The Swiss government has invested Swiss tax payer money into astronomy and space research, mathematics, biology, physics, chemistry, engineering, finance, psychology, meteorology, agriculture, business, and art. I call this a diversified investment portfolio. The outcome from this investment is products, companies, intellectual property, scientific publications, jobs, tourists, a long list of benefits that can be measured by money. The WIPO innovation index[11] lists all of these innovation inputs and their innovation outputs per country. Money going in, money coming

11 Global Innovation Index (GII), The World Intellectual Property Organization (WIPO), September 29, 2022

out. It's a wonderful tool where a government innovation enthusiast can get lost for hours. Trust me, I speak from experience.

In this first section, I will go through some of the tools you as a country or a state leader have at your disposal. The tools are important to explore, as they collectively give you ammunition against arguments such as "the government should not pick winners and losers," or "the government should just get out of the way."

Any intelligent person, if they would only pause for a minute and be honest with themselves, can see that government cannot just drop doing these things and "get out of the way." The government, by doing what it is supposed to do, is picking winners and losers every day. It's literally the job.

KEY TAKEAWAYS:

1. Major economic terms and ideals were established in the 18th, 19th, and 20th centuries.
2. The role of government in economy has been debated for centuries.
3. High government oversight is not blocking commercial or economic success.

LAWS AND REGULATIONS

The British Industrial revolution that ran from around 1760 to the early or mid-1800's was the kickoff of our modern technology industries. Throughout human history up until those decades, people of all ages excluding the very most upper-class elites worked long hours in poor conditions just to be able to feed and shelter themselves. This included children.

The Economic History Association, a US based body consisting of members of academia from around the world, has studied child labor during those years and how British law makers struggled to pass laws that would address the issue, but yet not stifle industrial growth.[12] The first labor law passed in Britain to protect children from poor working conditions, the Chimney Sweepers Act of 1788, attempted to improve the plight of what was called 'climbing boys.'

> *"Children who worked in homes were either apprentices, chimney sweeps, domestic servants, or assistants in the family business. As apprentices, children lived and worked with their master who established a workshop in his home or attached to the back of his cottage. The children received training in the trade instead of wages. Once*

12 Carolyn Tuttle, Child Labor during the British Industrial Revolution, the Economic History Association, January 2011

they became fairly skilled in the trade they became journeymen. By the time they reached the age of twenty-one, most could start their own business because they had become highly skilled masters.

Both parents and children considered this a fair arrangement unless the master was abusive. The infamous chimney sweeps, however, had apprenticeships considered especially harmful and exploitative. Boys as young as four would work for a master sweep who would send them up the narrow chimneys of British homes to scrape the soot off the sides."

Other luckier boys worked on farms looking after animals, planting seeds, pulling weeds and harvesting ripe crops. Girls would leave their homes at the age of twelve to become domestic servants, or would milk cows and care for chickens.

The Chimney Sweepers Act of 1788 stated that boys under eight years old should not be bound to these apprenticeships, and that all boys should be provided with adequate clothes and housing, and allowed to go to church on Sundays. The bill also tried to propose that Chimney Sweeping should be a licensed business but this clause was voted down. A penalty for keeping a boy under 8 years of age was to "*pay for every such Apprentice or Servant, so by him or her had, taken, employed, retained, or kept, any Sum not exceeding ten Pounds. nor less than five Pounds.*[13]" Five pounds in 1788 equals roughly £800 today or $US 925.00. This law changed very little, if anything in the ongoing practices of the industry.

The next legislation to regulate child labor, the Cotton Factories Act from 1819, mandated that no child under the age of nine was to be employed in cotton mills, with a maximum working day of twelve hours for all those under 16. While the chimney sweepers were mostly boys, textile workers were girls. Once again, the law had little effect. In fact, the number of underaged children working in Textiles and Dyeing continued to increase, reaching a peak in 1861 when 19% of workers in

13 Chimney Sweepers Act 1788, HMSO and the Queen's Printer for Scotland

British textile mills were children under 15 years of age, and throughout this period more than half were under 19.[14]

The industrial revolution was a busy time for British lawmakers. The Chimney Sweepers Act of 1788 coincides with the Slave Trade Act of 1788 which ended up being slightly more effective.

Towards the end of the 18th century in Britain, the slave trade was beginning to raise opposition. The so-called Zong massacre of 1781 fueled these sentiments. Zong was a British ship carrying slave "cargo" from Accra in West Africa (current day Ghana) to Jamaica. Leaving Accra with 442 enslaved Africans on board, the ship was grossly overcrowded with more than double the amount of slaves than the ship was built to carry. This was usual in those times. At some point during the journey, the ship ran out of drinking water and the captain solved this problem by throwing 130 slaves overboard.

The owner of Zong had taken out insurance against the cargo, and ensuing court cases to collect their insurance money revealed the detailed horrors of slave trade to the public. For example, the Chief Justice of the King's Bench in London reviewing the case compared the slaves to horses, seeing nothing wrong in the cargo owner trying to salvage most of his cargo by killing some portion of it. The trial of March of 1783 found that the insurance company was liable for the damages to the cargo, as legally, the enslaved people were the same as any other cargo. Two months later, the Chief Justice overturned the decision due to new evidence, but his statement that the enslaved were equivalent to horses remained the opinion of Britain's highest court.[15]

The Abolitionists were able to lobby for the Slave Trade Act of 1788 which limited the number of enslaved people carried on British vessels. It was the first British law to regulate the slave trade, and was successful in resulting in a decline in slave mortality on British ships.[16] Slavery was abolished everywhere in the British Empire in 1803.

---

14 Carolyn Tuttle, Child Labor during the British Industrial Revolution, the Economic History Association, January 2011

15 Zong slave ship trial, This Day in History (June 22), HISTORY, June 10,2020

16 Slave Trade Act 1788, Wikipedia

What does this have to do with anything we are trying to accomplish today? It doesn't sound like passing these laws made it any easier for businesses to generate profits and hire more people. This could be the case, but it did make life easier for enslaved people, and for those companies who did not base their business in exploiting others. Trying to regulate the use of children and other enslaved people for profit, and enforcing these laws has been a slow process which continues today.

While most countries have outlawed slavery, these same countries continue to enjoy the benefits of modern slavery in the form of inexpensive goods and services. Sometimes the western world's anti-slavery efforts are most successful when evaluated from the point of view that all businesses need to have a level playing field—no unfair advantage for one business over another—as opposed to the ethical view that slavery is wrong.

Modern anti-slavery government actions include tax audits into the black market targeting businesses who operate under the radar, laws banning the import of goods made with forced labor, and crackdowns on illegal immigrants working in agriculture, construction, and hospitality. Charges for the use of enslaved people for business profits are often linked to charges for money laundering. While no one would argue that the use of slaves is ethical or a good business practice, we tend to look at it as a violation to our fellow community members in terms of lost tax revenue and unfair advantage in the free market. More on this topic under the section on trade agreements.

A similar legislative approach could be shown about pollution control. President Richard Nixon signed the National Environmental Policy Act and created the Environment Protection Agency in 1970 but the business community is only just now beginning to embrace these ideas. It has taken a severe climate crisis for people to see what uncapped pollution does to our planet. Even though polluting the environment is rapidly beginning to be on par with slavery as a no-no, some still argue that polluting the environment is good for business, therefore good for the communities in terms of jobs and tax revenue.[17] Everyone would nonetheless agree it is

17 Palash Ghosh, Here's Why Investors Are Pouring Money Into Coal, Despite Its Bleak Future, Forbes, May 6, 2021

unfair to let one company pollute while others do not have this advantage (again, the argument of a level playing field).

By the way, legislating for ethics is *not* bad for business. History shows many cases where consumers successfully boycotted goods or services from countries that were behaving unethically. Growing up, we refused to buy South African fruit until apartheid was ended; I still avoid Israeli products as a private personal strike against the occupation of West Bank. Russia's economy is basically bust due to their unethical behavior in Ukraine. Consumers are showing their political power by hurting businesses. As a business owner, you should be happy your country has internationally recognized laws around ethical practices.

So, we've established that level playing field and ethical ways of working provide a sustainable environment for the private sector to thrive. Not all laws are trying to restrict businesses or create a level playing field. Some government acts have been instrumental for the very existence of an innovative, competitive business environment. The modern economy would not be what it is today if it wasn't for the Limited Liability Company, which came about in the following way.

The British ships carrying slaves and other goods on behalf of their kingdom, as well as the Portuguese and the Spanish ships for their kingdoms, were organized in various ways including as monopolies or under licenses granted by the king.

The British ship Zong that had taken insurance against the goods (people) it was carrying was organized by The Gregson Syndicate—Mr. William Gregson, his two sons, a son-in-law and two additional individuals. The members of this syndicate personally shared the profits, and were personally responsible for all of the losses. This explains why they were ready to sue their insurance company, risking the public outcry for the massacre they committed in order to receive the insurance money they felt was owed to the syndicate.

Very few individuals were wealthy enough (and also willing) to be taking such high personal risks in order to start commerce. In the early to mid-19th century the United States, Britain, and France recognized this as an unnecessary threshold to trade, and passed laws allowing for individual people to pool together and each person be financially liable for only the sum he had invested. These laws formed the idea of a Limited Liability Company, the LLC.

The first of such laws was passed in the State of New York in 1811, although the first actual LLC was not formed in the USA until 1977 in Wyoming. England passed the Joint Stock Companies Act in 1844 and the Limited Liability Act in 1855. The first Companies Act in Sweden was enacted 1848 under the influence of similar French laws.

By the final quarter of the 1800s most of Europe and most US States had limited liability laws in place which paved the way to large-risk enterprises in steel, energy, textiles, and transportation.[18] In each of these countries, governments proactively saw they needed to advance business and allow for more risk taking in order for the entire economy to benefit.

More recent landmark legislations that fuel private innovation are linked with technology transfer from publicly funded labs into the private market.

Before the Bayh-Dole Act of 1980 in the United States, any invention coming out of a government funded lab (this would be a major university, a national lab working directly under Department of Energy, Department of Defense and others, even any private company who received public funding), would be considered the intellectual property (IP) of the United States. The Bayh-Dole Act made it possible for those organizations to retain ownership of their own inventions, and license them out to third parties.

Perhaps the industry most benefitting from the Bayh-Dole Act is the pharmaceutical industry. Developing a new drug is very expensive, so introducing government-backed funding to small- and medium-sized drug companies bolstered innovation immediately. Over 200 new drugs and vaccines have been developed since the Bayh-Dole Act, including Hepatitis A Vaccine, Adenovirus Types 4 and 7 Vaccine, The Next Generation Diagnostic System, and Convulsant Antidote for Nerve Agent. Vaccines are in the pipeline for Malaria, Dengue, HIV, Ebola, MERS, Zika,

18 Limited liability, Wikipedia

Oldest existing companies in Europe.

Country	Company	Year
Albania	ALBTELECOM	1912
Andorra	ANDBANK	1930
Armenia	YEREVAN ARARAT BRANDY-WINE-VODKA FACTORY	1877
Austria	ST. PETER STIFTS KULINARIUM	803
Belarus	OLIVARIA BREWERY	1864
Belgium	AFFLIGEM BREWERY	1074
Bosnia and Herzegovina	SARAJEVSKA PIVARA	1864
Bulgaria	ARSENAL AD	1878
Croatia	KRALJEVICA SHIPYARD	1729
Cyprus	BANK OF CYPRUS	1899
Czech Republic	PIVOVAR BROUMOV	1348
Denmark	MUNKE MØLLE	1135
England	THE ROYAL MINT	886
Estonia	RAEAPTEEK	1422
Finland	FISKARS	1649
France	MONNAIE DE PARIS	864
Germany	STAFFELTER HOF WINERY	862
Greece	PIRAEUS BANK	1606
Hungary	ZWACK	1790
Iceland	ÍSLANDSPÓSTUR	1776
Ireland	SEAN'S BAR	900
Italy	MARINELLI BELL FOUNDRY	1040
Kosovo	MERIDIAN CORPORATION	1999
Latvia	CĒSU ALUS	1590
Liechtenstein	NATIONAL BANK OF LIECHTENSTEIN	1861
Lithuania	GUBERNIJA	1665
Luxembourg	MOUSEL	1511
Malta	HSBC BANK MALTA	1882
Moldova	NO DATA	No data
Monaco	NO DATA	No data
Montenegro	POŠTA CRNE GORE	1841
Netherlands	THE BRAND BREWERY	1340
North Macedonia	TUTUNSKI KOMBINAT PRILEP	1873
Norway	POSTEN NORGE	1647
Poland	BOCHNIA SALT MINE	1248
Portugal	CTT-CORREIOS DE PORTUGAL SA	1520
Romania	URSUS BREWERIES	1878
Scotland	BANK OF SCOTLAND	1695
Serbia	APATIN BREWERY	1756
Slovakia	KREMNICA MINT	1328
Slovenia	GOSTILNA GASTUŽ	1467
Spain	CASA DE GANADEROS	1218
Sweden	SKYLLBERGS BRUK	1346
Switzerland	GASTHOF STERNEN	1230
Turkey	ÇEMBERLITAŞ HAMAMI	1584
Ukraine	DROHOBYCH SALT PLANT	1250

Data source Business Financing Ltd. Image credit Petra Soderling.

The oldest existing companies in Europe.

and Ricin.[19] 70% of global biopharma's clinical trials take place in small companies who otherwise could not afford these programs.

The same year as the Bayh-Dole Act, 1980, Congress passed the Stevenson-Wydler Technology Innovation Act which required federal laboratories to actively participate in and budget for technology transfer activities so that nonfederal private entities could make use of and benefit from government-funded inventions. As a result, open the budget in any federal lab, be it the Defense Advanced Research Projects Agency (DARPA), National Aeronautics and Space Administration (NASA), or the Air Force Research Lab (AFRL), and you will see a line item for technology transfers from their lab to commercial use, from the public to the private sector.

I can't help but imagine what Milton Friedman might have felt about these two laws, both introduced by Democratic senators and signed into law by Jimmy Carter, a Democratic president. In his book Capitalism and Freedom,[20] Friedman did say,

> *"Government is essential both as a forum for determining the "rules of the game" and as an umpire to interpret and enforce the rules decided on. What the market does is to reduce greatly the range of issues that must be decided through political means, and thereby to minimize the extent to which government need participate directly in the game."*

Would he have considered opening public/federal IP to private companies as 'participating in the game?' The answer most likely lies in the microfilm archives from 1980-81 (and if you have access to those let me know). Remember I mentioned that Dr. Friedman served as President Reagan's economic advisor. Ronald Reagan took office one year after the Bayh-Dole Act was passed, launching a new economic direction they called 'Reaganomics'. Reaganomics, in line with Friedman's thesis,

---

19 Bayh-Dole Act (1980), Biotechnology Innovation Organization

20 David Kiron and Gregory Unruh, Business Needs a Safety Net, MIT Sloan Management Review, November 01, 2017

claimed that the less the government meddles with the private sector the better. So perhaps he would have liked to see those 200 drugs and vaccines invented in another way, or not at all.

Regulating child labor, slavery, pollution, intellectual property rights, marketing claims, consumer privacy, and bankruptcy are the governments' way of ensuring their countries have a fertile and fair working environment for industries to emerge and businesses to grow. This is important to recognize, not only as the right thing to do, but as the most profitable thing to do.

KEY TAKEAWAYS:

1. The legislative branch of government creates laws that enable private business to operate, grow, and renew.
2. The judicial branch of government creates a level playing field for all businesses and their customers.
3. Government is responsible for creating an environment of trust, transparency, and ethics.

INTEREST RATES AND TAXES

As the Covid-19 pandemic took over the world in 2020, many governments acted quickly to support their local economies. The Federal Reserve of the United States (The Fed)[21] allocated money to be lent to households, businesses, and state and local governments. They also cut what is called 'the target for the federal funds rate', also known as the overnight lending interest from bank to bank, to almost zero. The federal funds rate is an interest rate that is being followed by all other private lenders, so a zero percent federal funds rate brings down all other interest rates very rapidly. The Fed also put out forward looking statements on the future of interest rates so the economy can make investments relying on official prediction. "We are deploying these lending powers to an unprecedented extent [and] … will continue to use these powers forcefully, proactively, and aggressively until we are confident that we are solidly on the road to recovery," said Jerome Powell, the Chair of Federal Reserve, in April 2020.

The government's role in adjusting interest rates that keep the business environment stable is straightforward. When the interest rate is too high, businesses stop borrowing money. When businesses stop borrowing,

21 Eric Milstein and David Wessel, What did the Fed do in response to the COVID-19 crisis?, The Brookings Institution, December 17, 2021

they stop investing. This results in the economy slowing down. When the interest rate is too low, businesses overspend, the market heats up and eventually the economy finds itself in inflation.

Another Fed activity in supporting Covid-affected private businesses in 2020 was supporting the financial market to keep it operational. By purchasing massive amounts of securities, and lending money to securities firms, they pumped trillions of dollars into the financial market thus helping the market to support businesses during distress. In addition, the Fed also lent money directly to banks and employers.

In Europe's Eurozone, the European Central Bank (ECB) had a similar response[22]. The ECB kept their key interest rates at historically low levels so borrowing costs remained low. In fact, the overnight deposit interest in the Eurozone was even negative. The ECB also supported access to credit for firms and households, ensured short-term financial concerns did not prevent lending, increased banks' lending capacity, and as Christine Lagarde, the President of the ECB, said, to "Preserve financial stability through international cooperation."

This international cooperation that both the Fed and the ECB refer to means reactivating so-called swap lines. ECB website explains swap lines as follows:

> *"Central banks around the world hold reserves of currencies that are not their own. This is because their domestic banks also do business in these currencies, and thus sometimes require foreign-currency loans in the course of daily business.*
>
> *In times of great uncertainty, customers' demand for foreign currency assets can increase. If banks do not have enough foreign currency reserves on hand to meet increased demand, markets can become unstable. So central banks have established so-called currency swap lines. These swap lines let central banks of one country exchange their national currency reserves for those of the central bank of another country – thus ensuring that central banks can meet increased demand."*

22 Christine Lagarde, Our response to the coronavirus pandemic, European Central Bank

What this means is that, for example, the Bank of Japan holds a block of US dollars in case Japan-based businesses need dollars; the Bank of Canada holds a block of euros for those needing euros in Canada, and the Fed will hold Swiss Francs for any business needs in the USA that use Swiss currency. The Fed has swap lines in place with Canada, England, the Eurozone, Japan, and Switzerland. During 2020, the Fed also extended temporary swaps to the central banks of Australia, Brazil, Denmark, Korea, Mexico, New Zealand, Norway, Singapore, and Sweden.

The Fed's Covid strategy worked. Companies like Amazon, General Motors, AT&T and Boeing took on new, inexpensive debt or refinanced existing leverage creating a stronger bottom line. Amazon was particularly active which made Los Angeles Times write in May 2020 that "Amazon borrows $18.5 billion it doesn't need in new debt sale."[23] It is understandably the job of journalists to be the watchdogs over public spending, however in this case, the Fed's strategy was working exactly as intended. Private companies kept the financial markets active. Whether the private company uses that money to buy back their own shares (hence strengthening their own bottom line) rather than investing in their workforce or new product development is not for the Fed to decide. Strong companies form a strong economy as long as the other market dynamics function as intended.

The image below describes well how the federal funds target rate guides economic development. During times of recession, the Fed lowers the target rate, and the economy takes off. In the threat of the economy overheating, the Fed raises the interest rate again, cooling down the economy.

When it comes to taxes, the picture is more nuanced and can (more) easily flame an ideological debate. Companies pay taxes and other fees, use taxed services and employ people who then pay income tax. Governments then collect these taxes and re-invest the money back into society in various ways. In general, publicly paid infrastructure, education, and healthcare enable companies to grow and thrive. As we read in the

23 Molly Smith and Matt Day, Amazon borrows $18.5 billion it doesn't need in new debt sale, Bloomberg, May 10, 2021

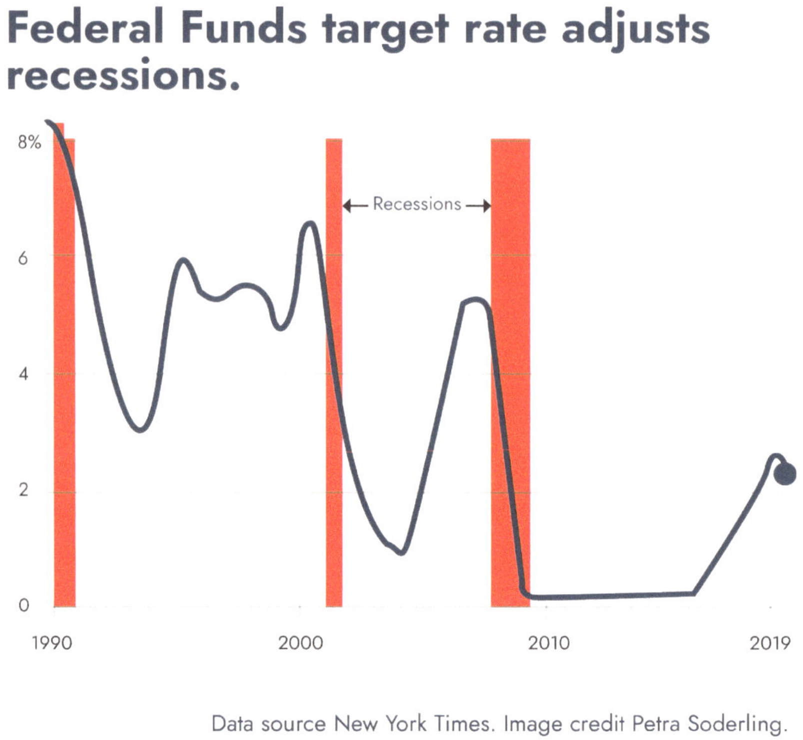

The target interest rate set by the Fed is an effective tool for adjusting recessions.

previous chapter, sometimes businesses even access intellectual property directly from a publicly-funded source.

The intent and the outcomes don't always match, however. In 2017, the US Congress passed the Tax Cuts and Jobs Act. At the time, President Donald J. Trump described it as a first step toward "slashing business taxes so employers can create jobs, raise wages, and dominate their competition around the world."[24] [25]A Stanford University study from 2019[26] looked

24 Dylan Walsh, The Mixed Impact of U.S. Corporate Tax Cuts, Stanford Graduate School of Business, October 28, 2019

25 William G. Gale, Did the 2017 tax cut—the Tax Cuts and Jobs Act—pay for itself?, The Brookings Institution, February 14, 2020

26 Rebecca Lester, Made in the USA? A Study of Firm Responses to Domestic Production Incentives, Journal of Accounting Research, Stanford Graduate School of Business, 2019

at a similar corporate tax cut decision from 2004 and found that those companies with international operations used the newly found funds to invest in automation in the US (hence reducing employment domestically) but did hire new people in other countries. They also invested in overseas acquisitions and—as we've just seen with Amazon—bought back their own shares. Only those US businesses who did not have any overseas activities created new jobs within the US. In light of this experience, only the third argument from Mr. Trump would be true; US businesses did dominate their competition around the world. Over a long term, the economy has other corrective actions, so automation and foreign IP will benefit the US taxpayer but not in the speed and magnitude intended.

Today, the federal corporate tax rate in the United States is 21% (2022). This is a flat fee on profits after deductions are made. It wasn't always so. In the past, the US federal corporate tax rate was divided into graduated taxable brackets based on taxable income. The higher the corporate income, the higher the tax percentage would be. Many state taxes still follow these graduated brackets. This is a way for these states to support smaller businesses allowing them to invest part of the earnings back into the company rather than pay it out in corporate tax.

In addition to corporate tax there are other taxes affecting business. Sales tax or value added tax (VAT) (a tax that is collected with a transaction by the seller and reported back to tax authorities) can make a product or service more or less attractive to the buyer. Many US and European states have different sales tax categories for food, books, public transportation, etc. This way the government can support businesses that are regarded as essential for the economy. Sales or VAT tax exemptions can also be used to guide behavior. As some examples of this, the state might want to encourage people to subscribe to clean energy, buy unprepared food, buy newspapers, or invest in farm equipment.

The income tax, a tax that is paid by an employee on the salary they receive, is a graduated tax form in most western countries. The US currently (2021) has seven income tax brackets, while most European countries have a sliding progression. Other countries mentioned in this book use one of these systems: India (seven brackets), Israel (seven brackets), Singapore (sliding progression), and South-Korea (sliding progression).[27] By keeping the employees income tax low, the government is assisting businesses in their personnel costs.

Interest rates and taxes are tools in policy makers' toolbox that can have large economic effects in a short amount of time. They are also tools that work best when launched together with a long-term strategy and combined with other, more surgical tools to produce a desired outcome.

KEY TAKEAWAYS:

1. Central banks regulate target rates in order to stabilize the economy.
2. National, regional, and local governments collect taxes to fund societies' needs.
3. National, regional and local governments use tax rates to guide behavior.

27 Tax Database Key Tax Rate Indicators, The Organisation for Economic Co-operation and Development (OECD), September 2021

TRADE AGREEMENTS

History is full of comical-sounding trade wars such as the opium war, the chicken war, the Cow War, the Pasta War and the Banana War. We also remember the Tulip Mania and the Railway Mania which were lessons on market bubbles, not to mention the Boston Tea Party.

The 21st century has its share of humor. In 2018, the US and China started a ping-pong match on raising tariffs on goods including fake teeth, flame-throwers, and cassette tapes. When the North American Free Trade Agreement (NAFTA) went into effect in 1994, US broom manufacturers blamed Mexico for flooding the US market with cheap brooms resulting in 200 broom workers losing their jobs. Maybe the trade wars weren't so funny after all.

On a more serious note, headlines like these should make CEOs' furrow their brows: "UK-France fishing reprisals threaten full post-Brexit trade war", "EU braces for steel quotas to end trade war with Biden", "US-China LNG deal a 'ceasefire' before next trade battle", "NAFTA and USMCA: The Next Stage of the Saga". Trade agreements, trade disagreements, and outright trade war? It's everywhere.

Against this aggressive backdrop, it may be surprising to learn that trade agreements were originally designed to enhance peace.

The World Trade Organization (WTO)[28] is an international body that is formed to set the rules of trade between countries. WTO trade

28 The World Trade Organization (WTO)

agreements intend to ensure that trade flows as smoothly, predictably and freely as possible. Much like its precursor the United Nations, it was established after World War Two to re-introduce harmony and peace among nations. WTO does not negotiate trade agreements directly, but it provides a framework for negotiations which usually aim to reduce or eliminate tariffs, quotas, and other restrictions.[29]

At the time of this writing, the European Union is not happy with how the United States handled their Inflation Protection Act from the summer of 2022. The EU claims that subsidies offered by the US government to US businesses and consumers on electricity, aviation, and hydrogen are not acceptable and harm international trade. Europeans threaten to take the dispute to the WTO, possibly together with South Korea and Japan. The plot thickens, as Europe, Korea, and Japan are all countries actively developing and selling similar, competing technologies.

In her 2016 book 'Shadow Courts: The Tribunals that Rule Global Trade'[30], investigative journalist Haley Sweetland Edwards studied WTO, international trade cases, and came to the conclusion that modern trade disputes are less about shipping goods across borders, and more about disagreements about moral and ethical rules. Issues like the environment, genetically modified food, dolphin-free tuna, or racial inequality are being discussed within trade agreements in conjunction with exports, imports, and GDP—topped off with a vision of what kind of world we want to live in.

This aspect of the trade negotiating and trade disputes sounds interesting. Let's dive deeper into it.

The world's largest trade agreement (2022) is the Regional Comprehensive Economic Partnership (RCEP), signed by 15 Asian countries including Australia, Brunei Darussalam, Cambodia, China, Indonesia, Japan, South Korea, Laos, Malaysia, Myanmar, New Zealand, the Philippines, Singapore, Thailand and Vietnam. This pact represents almost a third of the global population and over $26 trillion in gross domestic product. It is even bigger than the United States–Mexico–Canada Agreement (USMCA), formerly the North American Free Trade Agreement (NAFTA).

29 The World Trade Organization, Wikipedia
30 Shadow Courts: The Tribunals that Rule Global Trade, Amazon, September 6, 2016

The US and India, led at the time by conservative leaders Donald J. Trump and Narendra Modi respectively, decided not to join the RCEP, citing concerns about China having too strong a role. These concerns had to do with both national security as well as President Trump's 'America First' strategy which was seeking to bring overseas jobs back to America. This time, the domestic and geo-political situations were ranked higher than the ability to trade freely with this group of nations.

Renegotiating NAFTA into the USMCA gives us another recent look into the why and how nations make trade agreements and the arguments they present. The North American Free Trade Agreement (NAFTA) was signed in 1994 to eliminate tariffs between Mexico, United States, and Canada mostly in agriculture, textiles, and automobile manufacturing. Basically, NAFTA was a way to legally bring in Mexico's lower-cost workers to benefit its northern neighbors. This allowed for US automakers and consumer electronics companies to set up factories in Mexico and compete against products made in Asia. NAFTA members also enjoyed tariff-free agricultural products such as avocados and fruit from Mexico, grains and meat from the USA, and lumber from Canada.

It has been estimated Mexico gained hundreds of thousands of new jobs and enjoyed a lower consumer goods prices. This set off critics in the USA who argued that these were jobs that should be brought back to America. In 2018, NAFTA was renegotiated into USMCA with small adjustments. The main change was increasing regulations on the natural environment and the working environment. Also, greater incentives were added for automobile production in the U.S. and more access to Canada's dairy market in the other regions.[31]

Whether transforming NAFTA into USMCA was a political rather than economical deal is not for me to argue. However, it is undeniable that trade agreements, or lack thereof, can be tools for both domestic politics as well as international diplomacy. Companies and innovation may or may not benefit immediately.

It is notable that the countries with most trade agreements in place are the same countries that thrive economically, have the best education systems, and are most innovative.

---

31 United States–Mexico–Canada Agreement, Wikipedia

Countries with most trade agreements 2021.

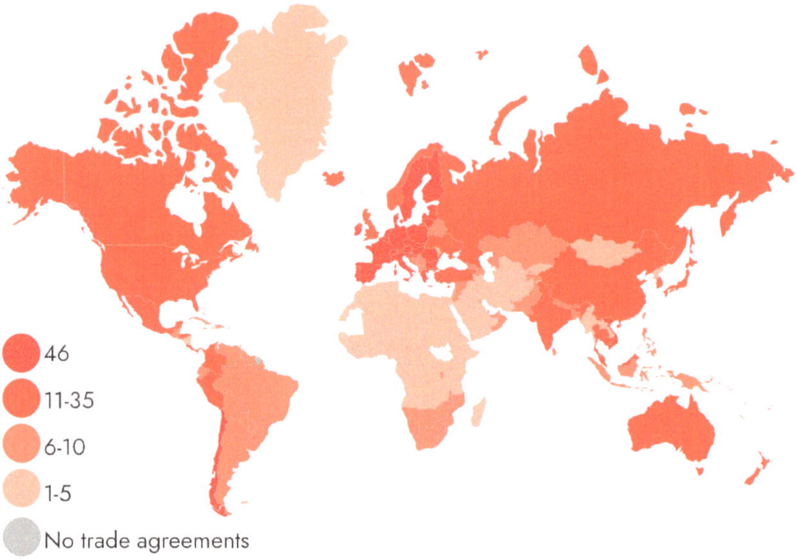

Data source WTO. Image credit Petra Soderling.

IMAGE 4:Countries with most and least trade agreements

Despite the United States being the European Union's largest trading partner, there is no trade agreement in place. Between these two regions, the EU and the US have the largest bilateral trade in the world, and either the EU or the US is the largest trade and investment partner of almost every other country in the global economy. Combined, these territories account for over 40% of the world's GDP and trade.[32]

Throughout history, a trade agreement for these two players has been on the drawing board. The last time it was brought to the table was in 2018. The Transatlantic Trade and Investment Partnership (TTIP), according to the EU, would have boosted the EU's economy by €120 billion, the US economy by €90 billion and the rest of the world by €100 billion. The

---

32 EU trade relations with the United States. Facts, figures and latest developments, European Commission

Office of the US Trade Representative[33] saw it similarly; however, there was also much criticism both from the EU members as well as American stakeholders. President Trump backed out of the negotiations in 2019. Commerce between these regions continues to flow even without a trade agreement in place.

As you can see, trade is often more than trade. It is a way to signal friendship, hostility, values and strategies. Sometimes this is good for your industry in the short-term; sometimes businesses need to wait for a longer time to benefit from your government's international contracts. When our governments decide to open up new business bridges across borders it gives us hope and optimism, and trust that our government have our backs.

KEY TAKEAWAYS:

1. Countries and regions form trade agreements to support and protect their local businesses.
2. Countries and regions form trade agreements to create geopolitical pressure.
3. The most advanced economies have the most trade agreements.

33 Transatlantic Trade and Investment Partnership (T-TIP), Office of the United States Trade Representative

IMMIGRATION AND WAGES

"The Caravan is coming, build the wall".

If you live in the US, you have heard this line. The 'caravan' refers to a flood of immigrants from Central and South America heading to the southern US border hoping to get into the US, and thus into its labor force. These migrants are motivated both by fear and poverty in their own countries, as well as the lure of a better life in the United States.

The conversation around immigrants crossing the southern US border is twofold. On one hand, the US economy needs these workers for jobs that the existing population of Americans isn't willing to do. On the other hand, having a class of workers on the border of legal status brings down wages and creates an unfair advantage for those businesses choosing to employ migrant workers.

The Center for Immigration Studies published a report[34] on the effect of immigration to low-skilled labor markets. It found that in general—in all jobs nationwide in America—a one percent increase in the immigrant composition of an individual occupation reduces the weekly wages of natives in the same occupation by about 0.5 percent. Since roughly ten percent of the labor force at the time of this study was composed of immigrants, these findings suggest that immigration may reduce the wages of the average native worker by perhaps five percent.

In low-skilled occupations where workers have only a high school diploma or less, the effects of immigration are much stronger. For the twenty-three

34 Steven A. Camarota, The Wages of Immigration: The Effect on the Low-Skilled Labor Market, The Center for Immigration Studies, January 1, 1998

percent of natives employed in these occupations (about twenty-five million workers), a one percent increase in the immigrant composition of their occupation reduces wages by 0.8 percent. Since these occupations are fifteen percent immigrant, this suggests that immigration may reduce the wages of the average native in a low-skilled occupation by perhaps twelve percent, or $1,915 per year (expressed in 1998 wages when the study was published).

These numbers are purely statistical and imply natives and immigrants in low and high earning jobs are competing in the same job market. The more recent 2017 Cato Institute paper[35] disputes this view and claims "restricting immigration will not have a substantial positive impact on native wages, at least in real terms."

Instead, the Cato Institute argues that if immigration was further restricted, two things would follow. First, certain low-skilled jobs would be automated. Automation would have a number of other economic consequences, but it certainly would not raise wages in those same low-skilled jobs. Second, the number of high-skilled women in the workforce would decline, as there would no longer be immigrant-dependent low-cost childcare available. Following this line of thought, restricting immigration could result in automated work places that offer jobs to a few men. The rest of the men and the women currently working in low-skilled jobs would either become unemployed or train for higher paid jobs.

We can see how different immigration policies have affected the economy in two countries who recently took different directions: Canada and the United Kingdom (UK).

The UK decided to leave the European Union (EU) because it was seen as bureaucratic, expensive, and with it came open borders for immigrants from all over Europe and elsewhere. In his paper 'Immigration and the UK Economy after Brexit'[36], Jonathan Portes looks at how immigration policies and migration flows have changed starting from before the EU referendum in 2016 and up until early 2021 when the UK passed a new post-Brexit migration policy.

35 Alan de Brauw, Does Immigration Reduce Wages?, CATO Journal, Fall 2017

36 Jonathan Portes, Immigration and the UK Economy after Brexit, Social Science Research Network, 12 Jun 2021

When the EU expanded to the east in 2004 and 2011, the UK started to see an influx of migrants from old but especially new EU countries. In 2019, one in five EU citizens who had migrated within the EU lived in the UK. Portes' paper shows a wealth of evidence how this pool of migrants in the UK have directly benefited both the growth of GDP and GDP per capita. Two interesting data points include that "a one percentage point increase in the share of immigrants within a UK local authority is associated with an almost three percentage point increase in productivity" and that "the marginal migrant is around 2.5 times as productive as a UK-born worker."

After the referendum for the UK to withdraw from the EU won, immigration dropped dramatically. Because of Covid we don't have reliable statistics on what has actually happened since 2019, but Portes has projected the numbers of the positive impact migrants have had to the UK economy into the government's estimate on future immigration numbers.

> *"In our paper for the OXREP special issue on Brexit in 2017, Portes and Forte (2017a), we drew on two papers that used cross country evidence to estimate such impacts [Boubtane, Dumont and Rault, 2015; Jaumotte, Koloskova, and Saxena, 2016] and incorporated their estimates into our assessment of the impacts of post-Brexit changes to immigration. This led to much larger negative impacts, of between 3 and 8% to GDP, and 1 to 5% to per-capita GDP."*

In other words, in their pursuit to get rid of immigration, UK voters will be looking at a dramatic decline in their economic growth.

Canada, on the other hand, has been implementing a different immigration policy.

Long before the US put roadblocks on the H-1B visa type[37], which is often given to Asian engineers who entered the country to work for tech companies, Canada was attracting highly educated international talent. In 2017 Canada's population had the fifth-most foreign-born people in the world[38].

37 Roy Maurer, Trump Suspends New H-1B Visas Through 2020, The Society for Human Resource Management, June 23, 2020

38 Amelia Cheatham, What Is Canada's Immigration Policy?, The Council on Foreign Relations (CFR), February 9, 2022

Canada started to embrace multiculturalism in the late 1960s and early 1970s. Followed by a point-based, immigrant-friendly legislation in Ottawa, the Canadian government issued policies in 1971 and 1976 that committed the country to refugees and immigrants specifically to support cultural, economic, and social objectives.

Today, 75% of Canada's population growth comes from immigrants, and six in ten immigrants are selected for their positive economic impact. Between 2019 and 2021, Canada invited over 300,000 highly-educated immigrants each year, resulting in an innovation pulse that represents 1% of the Canadian population.[39]

How the two different approaches, UK's and Canada's, will eventually impact the countries' economies will remain to be seen. There continue to be conflicting narratives[40] on the impact of immigration to any economy. What is certain, though, is that all governments try to help their economies to thrive by ensuring the ratio of workers to retirees stays on a healthy level, new entrants will not be a burden on the economy, women will be able to join the workforce, and that businesses will be able to hire skilled workers. Immigration is not a silver bullet, but it could act as a short-term pressure valve and a long-term building block if any administration chooses to do so.

Followed by the passing of the CHIPS and Science Act in the Fall of 2022, a law that encourages the development of semiconductors and related technologies in America, the Senate Commerce Committee held a hearing on how to secure U.S. leadership in this area.[41] Witnesses in the hearing ranged from universities to private technology companies. While many important topics came up in the two-hour hearing, one subject was constant: Talent. Universities, Research Labs and private companies developing Artificial Intelligence (AI) solutions, blockchain technologies, designing and running High Performance Computers (HPC), and creating Quantum computers and solutions, they all consistently said that lack of educated people is the biggest obstacle for creating new innovation.

39 Infographic: Immigration and Economic Growth, The Government of Canada

40 Economic impact of immigration to Canada, Wikipedia

41 Securing U.S. Leadership in Emerging Compute Technologies, U.S. Senate Committees, September 29, 2022

Dr. Nancy Allbritton, Dean of the Frank and Julie Jungers Dean College of Engineering, University of Washington voiced this on behalf of all of the witnesses as follows:

> *"We need all the talent we can get. If you look at any university in the US you can see that the talent comes from across the world. Many of the companies we're talking about, their founders are not from the US but they have found home here. So being an open and welcoming space gives us a technological advantage that we can capitalize so that we have a brain-collective in the US. We'll be richer, more diverse, and more productive as a result."*

Most countries, or should I say most successful countries, regulate their minimum wage.[42] Much like the debate around immigration, the arguments for and against minimum wage will voice claims such as "minimum wage ensures employees are being paid a living wage and they can contribute to the economy with sufficient purchasing power", and "minimum wage puts an unnecessary government-regulated burden to companies already struggling with high operating costs." Either claim is difficult to prove right or wrong on its own, as most issues in the economy are deeply interwoven with other dynamics. The economy is like nature, where everything is connected. Humans can only increase or decrease the amount of pressure on one part of the ecosystem; in the short term, it may appear to result in the desired outcome but at some point adjustments are needed as there are bound to be unintended consequences somewhere else.

The Nordic countries of Iceland, Norway, Sweden, Finland and Denmark do not have a statutory minimum wage. They achieve the same goal via centralized, union-led collective agreements.[43] Collective agreements are negotiated amongst the government representative, the employees union, and the employers union. These collective agreements are binding not just for companies that are members of this union, but for all companies operating in the same field, and they often last for years.

42 Minimum Wage by Country 2022, World Population Review

43 Minimum wages in Nordic countries, International Labour Organization (ILO)

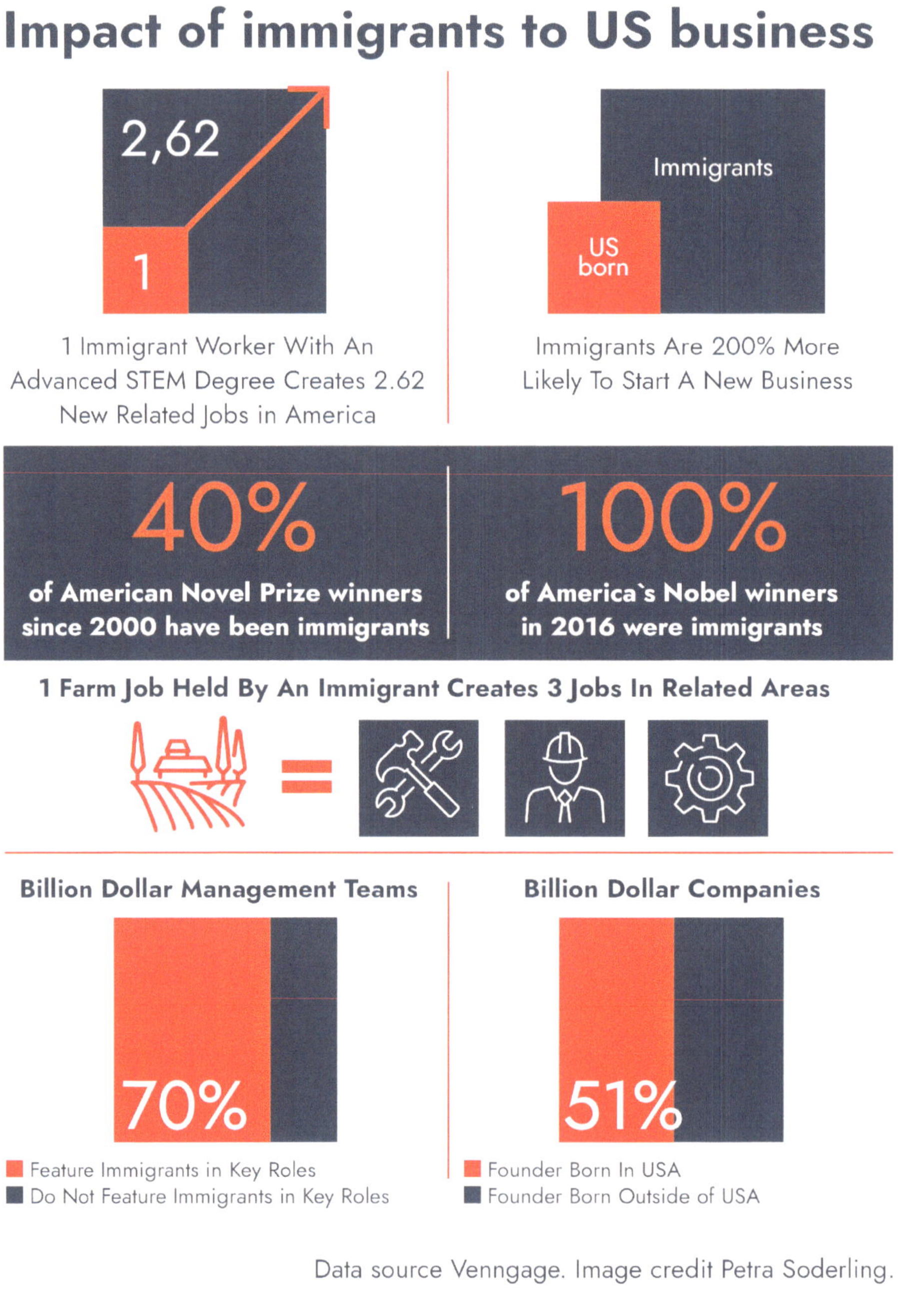

Immigrants have a positive impact to America's economy

Understandably, there are many non-unionized companies who feel this is a deeply unfair system, but there are just as many companies in the minimum wage countries who have similar sentiments. At least in the collective system employees have a say, and the wage range is not "dictated" by the government. On the other hand, these wage norms are much higher than the minimum wage so the employers may not be so happy. Again, coming back to the financial success of the "free economies" versus the "highly regulated economies", the Nordic model may seem expensive and unfair, but you cannot say that it has not been working for their economies.

The argument supporting the minimum wage is for employees to be able to live and work, educate their children and contribute to society.

In most countries, the minimum wage is not achieving this set of goals, so additional benefits are being offered. Rent control, or rent regulation is one of these. Rent regulation can include things like limiting the amount of rent a landlord may collect, controlling when and how landlords can evict families, and obligations for landlords to maintain properties. The purpose here is to make it hard for the so-called slumlords to play games on working people's ability to live a meaningful life. In the US, rent control is often maintained on a local level and varies much from state to state. According to the OECD, rent control in some form is available in 13 countries.[44]

Other ways to ensure employees are able to make ends meet and even have time to create side hustles and other innovative ideas include basic income, universal basic income, child support programs, elderly and other caregiver support, education support, food aid, and healthcare programs.

Immigration, wages, and wage-enhancing subsidies are some of the tools in your portfolio for making sure you will have a high performing, healthy and happy workforce that will generate the economic wealth your country or state is looking for. Employees and tax payers are rational and need to know how and where to invest their time and money so their children will have the maximum opportunity and freedom.

44 PH6.1 Rental Regulation, The Organisation for Economic Co-operation and Development (OECD)

KEY TAKEAWAYS:

1. Low-skilled immigrants enable native population to work at higher-paying jobs.
2. High-skilled immigrants create innovation, businesses, and jobs in their new country.
3. Government needs to ensure living income to families either by regulating the market or by offering subsidies directly to families.

INFRASTRUCTURE

On November 15, 2021, President Joe Biden signed into law a $1.2 trillion bill called the Infrastructure Investment and Jobs Act (IIJA).[45] As we will learn in this chapter, this law and budget was a long time coming. The bipartisan law has 2,702 pages, it passed by a 228 to 206 vote, and is said to finally modernize America's aging infrastructure to be on par with the rest of the world. President Biden said in the signing ceremony, "Generations from now, people will look back and know this is when America won the economic competition for the 21st century." When a politician says it, you may want to take it with a pinch of salt. But he was not alone. The Brookings Institution[46], an independent and non-governmental think tank, called IIJA "generation defining" and more meaningful than the New Deal from 1933. CNBC[47] said this bill "spurs hopes of a small business boom"; workers unions welcomed it; large corporations welcomed it. In June of 2022, another bill passed that will bring even more power to IIJA, the Inflation Reduction Act, that includes subsidies for clean energy.

---

45 Brian Naylor and Deirdre Walsh, Biden signs the $1 trillion bipartisan infrastructure bill into law, National Public Radio, November 15, 2021

46 Adie Tomer, Caroline George, Joseph W. Kane, and Andrew Bourne, America has an infrastructure bill. What happens next?, The Brookings Institution, November 9, 2021

47 Mikaela Cohen, Biden infrastructure bill has big business backing. Here is how small business feels, CNBC , August 6, 2021

This Infrastructure bill includes a large amount of funds allocated to transportation on land, sea, and air, as well as money for broadband internet, clean water and electricity, as seen in table below.

Fund allocation in the Infrastructure Investments and Jobs Ad, US dollars

Transportation	Roads, bridges, and other major projects	**$110 billion**
	Passenger and freight rail	**$66 billion**
	Public transit	**$39 billion**
	Airports	**$25 billion**
	Port infrastructure	**$17 billion**
	Transportation safety programs	**$11 billion,**
	Electric vehicles	**$7.5 billion**
	Zero and low-emission buses and ferries	**$7.5 billion**
	Revitalization of communities	**$1 billion**
Other infrastructure	Broadband	**$65 billion**
	Power infrastructure	**$73 billion**
	Clean drinking water	**$55 billion**
	Resilience and Western water storage	**$50 billion**
	Removal of pollution from water and soil	**$21 billion**

Image credit Petra Soderling.

How the budget associated with Infrastructure Investment and Jobs Act gets allocated, US dollars

This what the numbers look like on a pie chart:

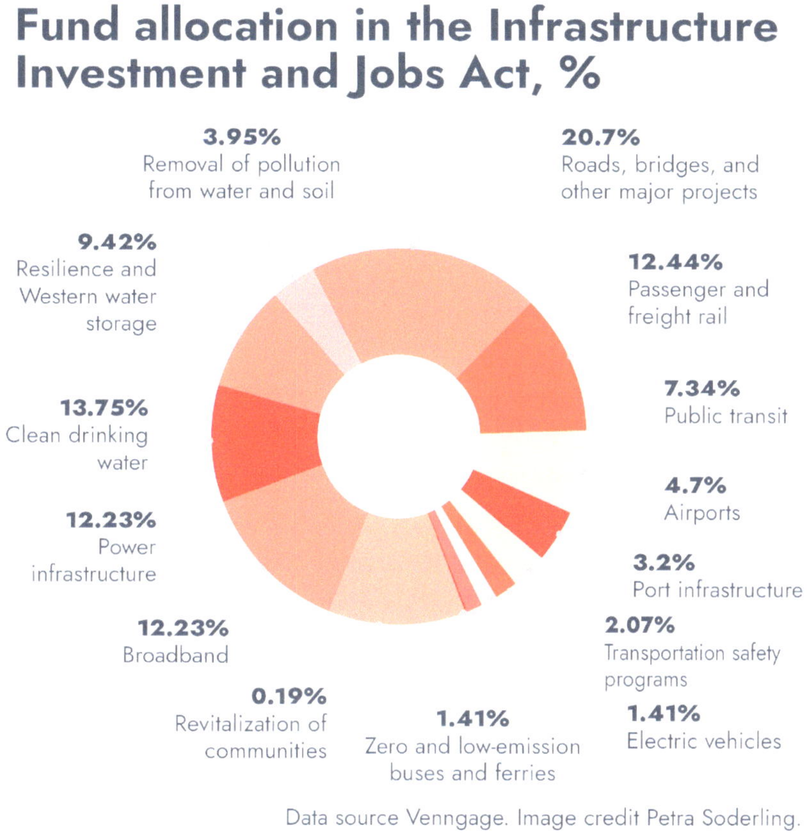

How the budget associated with Infrastructure Investment and Jobs Act gets allocated, percentages

Combined with other economic improvements by the administration, the White House estimates this will create 1.5 million new jobs per year for the next 10 years.[48] These new jobs come from direct implementation of these improvement plans.

What is not directly factored in these numbers though, is the massive improvement in the operating environment for US businesses. Better

48 Barbara Sprunt, Here's what's included in the bipartisan infrastructure law, National Public Radio, June 24, 2021

passenger and cargo transportation infrastructure, a more reliable energy grid, and accessible broadband directly make it easier, faster, and cheaper for companies to operate. Ensuring every American is able to use high bandwidth internet connectivity from their homes not only enables current work from home scenarios, but also allows children and students to improve their learning experience. Well-educated future generations are essential for growing any economy.

How does the US stack up against competition?

In 2018, China invests a whopping 5.57% of their GDP in infrastructure.[49] In contrast, before the Infrastructure Investment and Jobs Act, the USA invested only 0.52% of its GDP into infrastructure. This is less than 10% of what China is investing.

China's investments are part of the China 2025 plan, which aims for the country to achieve 70% self-sufficiency in high-tech industries by 2025, and to dominate the global markets by 2049 (the hundredth anniversary of the People's Republic of China). China has realized that infrastructure is critical for the energy- and water-heavy high-tech industry.

Infrastructure investment dollars and percentages are not the only criteria for governments to help their businesses. How effectively this money is allocated, distributed, and spent is being examined under the term 'governance'. The Global Infrastructure Hub[50] publishes an annual Infrastructure Compass that ranks nations in infrastructure governance among other parameters. The other parameters in this Infrastructure compass are Regulatory, Permits, Planning, Procurement, Activity, Funding, and Financial.

For 2020, the top 5 countries for infrastructure governance were Singapore, Denmark, Netherlands, Canada, and Austria. Other countries we'll discuss in this book include South-Korea (#13), and Finland (#14). China is #32 and on its way up. The US is at #36[51] and by 2020 had been going down in the charts for infrastructure governance, but also in planning and permitting. On the bright side, the US had gone

49 Global investments on the construction and maintenance of infrastructure as share of GDP in 2020, by selected countries, Statista Research Department, May 2022

50 These 5 countries lead the world in infrastructure governance, The Global Infrastructure Hub (GI Hub), October 1, 2020

51 The Global Infrastructure Hub (GI Hub)

up in regulatory framework and procurement. The Funding indicator in this Infrastructure Compass measures the ability for a country to fund infrastructure. The United States has always been high in that ranking as there is so much *potential* for public funding.

But what does this matter? How exactly does infrastructure grow the economy ?

A McKinsey report 'Four ways governments can get the most out of their infrastructure projects' from 2020[52] outlines that globally we spend more than US$2.5 trillion a year on infrastructure, but that US$3.7 trillion would be needed every year from today until 2035 just to keep up with the GDP growth we have estimated that we need. This is important because the GDP growth estimates are being used to allocate many other resources that affect people's lives every day.

Let me do the math for you: The world is missing US$1.2 trillion every year from its infrastructure investments if the goal were to just be able to keep the ball rolling. We would need Biden's once-in-a-lifetime-five-year-plan every year, and still we would not have made any progress. That's how much infrastructure debt there is.

To be blunt, economists are projecting GDP growth based on the goodwill of budget creators. If countries don't spend more, they will simply not grow their economies. Instead, they will fall behind.

As we've established however, there are big differences from country to country. Even though we live in a globalized economy, that does not mean countries are not in fierce competition with each other. Quite the opposite.

McKinsey continues to demonstrate that infrastructure's socioeconomic rate of return is around 20%. Every dollar invested increases the GDP by $1.20 in the long run.

The smart governance that Singapore, Denmark, Netherlands, and other successful economies practice include things like

1. developing projects with tangible, quantifiable benefits. Consider both the direct impact of an infrastructure project and the

52 Aaron Bielenberg, James Williams, and Jonathan Woetzel, Four ways governments can get the most out of their infrastructure projects, McKinsey & Company, January 6, 2020

indirect effects on the rest of the economy and society, rather than go after catering for one big current or potential future local employer, or a port infrastructure that helps seasonal cargo.

2. improve the coordination of infrastructure investments to account for network effects. If one regional airport is a bottleneck, consider building new airports instead of improving the logistics in that one airport.
3. consider the effects of infrastructure investments on social equity to ensure that the potential economic benefits are inclusive, cross income levels, urban and rural areas, and racial and ethnic lines.
4. and unlock long-term capital. Governments can work with investors that provide long-term capital to ensure that debt maturities align with time frames, taking into account when the benefits of the infrastructure are realized.

McKinsey said every dollar spent in public infrastructure adds $1.20 to GDP. In addition to this national benefit, on a project and local basis the return of a public infrastructure project could be as much as 45%[53]. This was calculated in a Josh Bivens report published by the Economic Policy Institute[54] where he proved the following outcomes.

- Investments in public capital have significant positive impacts on private-sector productivity, with estimated rates of return ranging from 15 percent to upwards of 45 percent. The preferred estimate is 30 percent, which coincidentally is roughly equivalent to the rate of return on investment in information and communications technology.
- A consistent research finding is that public capital offers a higher rate of return than most forms of private capital.
- Increasing public investment by just under $250 billion per year on average for the next 10 years could increase American gross domestic product by a greater percentage every year—producing

53 Investments in Infrastructure Bring High Returns, Grading The States
54 Josh Bivens, Public investment, The Economic Policy Institute, April 18, 2012

a GDP that is 0.9 to 2.8 percent higher in 2021 than it otherwise would have been. A significant fraction (40–75 percent) of the plan's budgetary cost could be essentially self-financing.

- Public investment has benefits that extend beyond simply increasing measured GDP, as it also offers benefits that are more broadly shared by all Americans.
- Public investment produces benefits that cannot be measured, such as safer water and cleaner air.

Public investment is at least as productive as private investment, and several strands of the research seem to indicate that it is substantially more productive.

Rates of Return of Public Investments

Bom and Ligthart (2008)	**17%**
De Haan and Jong-A-Ping (2008)	**30%**
Ascheaur (1989) and Isaksson's (2007)	**45%**

Economic Policy Institute report buy By Josh Bivens · April 18, 2012

Data source Economic Policy Institute. Image credit Petra Soderling.

Public Investments yield good rates of return

So how are these figures calculated? While there are several input-output models that address this question, Isabelle Cohen, Thomas Freiling, and Eric Robinson from the College of William & Mary, Thomas Jefferson Program in Public Policy in Williamsburg, Virginia[55], have potentially the best model that represents both short and long term effects.

In order to estimate the structural relationship between economic activity and public infrastructure investment, they use the vector

55 Isabelle Cohen, Thomas Freiling and Eric Robinson, The Economic Impact and Financing of Infrastructure Spending, College of William & Mary, 2012

autoregression (VAR) method explained in a paper by Alfredo Pereira, 'Is All Public Capital Created Equal?'[56].

The VAR method gives a numerical value that captures the dynamic effects that GDP and investment spending have on one another. The research team then adjusts this sensitivity for changes in the stock of different types of infrastructure (highways and streets; transportation and power; sewer and water; health, educational, office, public safety buildings; and conservation, development and nonmilitary equipment) which allows them to calculate the long-term permanent effect of infrastructure investment on GDP.

This image depicts the marginal product calculation for each type of public investment. Marginal product is a term to describe the overall economic output (GDP) that results from one initial dollar of spending in each area, in this example below over a twelve-year period.

Marginal Product Formula

$$MP_{z} = \varepsilon^{GDP}_{Investment_{z}} \; x \; \frac{\sum_{i=1998}^{2010} GDP_{i}}{\sum_{i=1998}^{2010} \Delta \, Net\ Capital\ Stock_{iz-(i-1)z}}$$

Data source Alfredo Pereira. Image credit Petra Soderling.

Marginal Product Formula calculates the economic output of investments over time

What the team discovered after creating this formula is that aggregate public investment in the selected five types of infrastructure is estimated to result in a marginal product of $3.21. In other words a one dollar investment yields $3.21 in GDP over 20 years. Transportation and power

56 Alfredo M. Pereira, Is All Public Capital Created Equal?, The Review of Economics and Statistics, MIT Press Direct, August 01, 2000

provided the largest positive economic impact with one dollar yielding $14 in GDP, while highways and streets were the lowest, with a 1 to 1.15 ratio.

While these mathematical formulas exist to calculate the return of a public infrastructure investment, this is rarely done as the time horizons are so long and variance can be large. We can try to test the theory in practice in projects that are large in size but narrow in focus.

Megaprojects are clearly defined projects, often planned for years if not decades in advance. Once up and running they provide a tremendous shift in economic growth. Public megaprojects[57] (projects that cost over US$1 billion) include several infrastructure projects.

Finland has relied on nuclear power since 1977 when the country's first plant started producing electricity. Since then, Finland has built four more reactors, the last of which is Olkiluoto 3. The 1600 MW Olkiluoto 3nuclear power plant was a 22-year megaproject and was finalized in January 2022. It cost €11 billion (US$12.5 billion) to build. The long-term Levelized Cost of Electricity (LCOE) target for all three plants is €30/MWh (US$34/MWh). The target was that, once Olkiluoto 3 became fully operational, Finland would have 40% of its energy from nuclear sources, and would become fully energy independent. In September 2022, the reactor achieved its full design output power. At the time of writing this, Finland still hasn't declared full energy independence, but the value of having a continuous stream of clean energy that will not be dependent on political decisions of other countries is clear for the country's population and businesses.

Another nuclear project on the global megaproject list is the Fukushima Daiichi nuclear disaster cleanup in Japan. The work aims to limit radioactive contamination from the three nuclear reactors involved in the 2011 earthquake and tsunami and consequent destruction of the power plant. The final price tag is not yet known, but in 2016, Japan's Ministry of Economy, Trade and Industry estimated the total cost will reach ¥21.5 trillion (US$187 billion). What this money will buy for Japanese businesses is again quite impossible to calculate, but for the Japanese government not to engage in a cleanup would be irresponsible.

---

57 List of megaprojects, Wikipedia

Closer to home, the Interstate Highway system that began in 1956, and was completed 62 years later, cost US taxpayers $500 billion. The California high speed rail between Los Angeles and San Francisco is budgeted to cost $100 billion, while the Sound Transit 3, a light rail and bus system in Seattle, has a budget of $50 billion. These and other projects in ports, airports and cities will undoubtedly directly benefit businesses operating locally, regionally, nationally and globally. $14 in outcome to each dollar invested, as we learned.

Despite our two nuclear megaproject examples having a bit of a negative vibe in this story, typically these and other energy investments—be they megaprojects or just your ordinary infrastructure investments—have the obvious capital return as well as positive fringe effects for decades to come. We can only imagine what the US economy would look like if the Interstate Highway system wasn't built, or how the West Coast would have developed without those majestical public transportation systems.

Public investments, especially in infrastructure, are a tested and proven way to increase a country's competitiveness. With better returns than from private investments, there is no mathematical reason to skip building public infrastructure.

KEY TAKEAWAYS:

1. There is a huge global infrastructure debt that is holding back economic development everywhere.
2. Public infrastructure investments have a better return than most private investments.
3. By investing in public infrastructure, countries can see immediate short-term and sustainable long-term economic benefits.

WAR AND PEACE

The United Nations (UN) was founded on October 24, 1945. A few months earlier, representatives from fifty countries had gathered in San Francisco to discuss how to prevent something like the World War II from ever happening again.[58] Tools in their toolbox included then (as now) international diplomacy, peace and security keeping, humanitarian assistance, defining and protecting human rights, and upholding international law.

The UN's peacekeeping operations are prepared by its Security Council and approved by the UN General Assembly. In the General Assembly each country has one vote, and for important matters such as a peacekeeping operation a two-thirds majority is needed. The UN does not have an army or equipment of its own, but member states are asked to provide military and police assistance. Some of this is compensated back to the nations.

For the fiscal year July 1, 2021 through June 30, 2022, the UN peacekeeping budget was $6.38 billion and it financed ten peacekeeping operations around the world. This is less than half of one percent of the military expenditures globally, which is estimated to be at $1,981 billion in 2020.[59] 39% of the global defense budget comes from the United States of America with China being second at 13%. In light of these numbers you could say that investing in war is much more popular than investing

58 History of the United Nations, United Nations
59 How We Are Funded, United Nations Peacekeeping

in peace, although the quote attributed to both Plato and Sun Tzu 'Si vis pacem, para bellum' (if you want peace, prepare for war) suggests that all military spending is intended for peacekeeping only.

Being the world's largest defense spender, the US also funds the majority of the UN peacekeeping operations (28%), followed by China (15%) and Japan (9%). The remaining budget comes from Germany, UK, France, Italy, Russia, Canada, and South Korea in this order, and the rest of the 193 members of the United Nations.

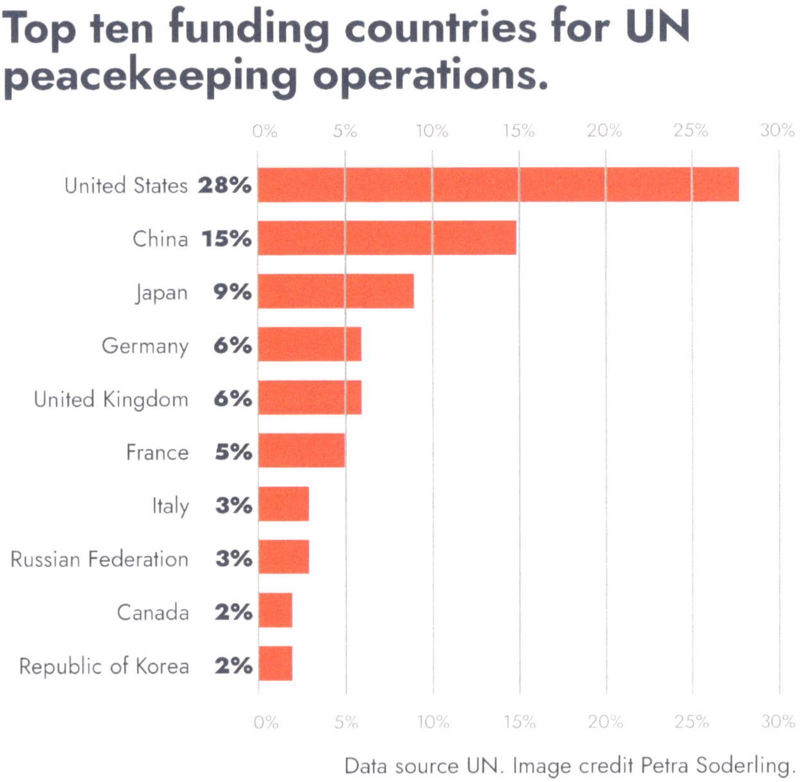

Top ten funders for United Nations peacekeeping operations

So, what does UN peacekeeping money buy?

While there are general guidelines and strict accounting principles applied to all peacekeeping operations,[60] for our purposes let's dissect a

60 Administrative and Budgetary Committee (Fifth Committee), United Nations

few ongoing operations to get a closer look. The most recent numbers for all of the UN's peacekeeping operations are found from a document called 'Approved resources for peacekeeping operations for the period from 1 July 2020 to 30 June 2021.' If you are interested in a bit of research, you will be able to calculate the combined investments from all of the peacekeeping operations into any single category, for example, Information and Communications Technology (ICT).

We'll look at three operations; The United Nations Multidimensional Integrated Stabilization Mission in Mali (MINUSMA), The United Nations Multidimensional Integrated Stabilization Mission in the Central African Republic (MINUSCA), and UNSOS, a UN Special Operation in Somalia. The full gross spending in 2020-21 for these operations was US$1.18 billion in Mali, US$1.12 billion in Central Africa, and US$550 million in Somalia.

Let's start in Mali. MINUSMA was established in 2013 and it deploys over 18,000 troops, military personnel, and police mainly from other African countries.[61] Its annual budget of over a billion dollars is roughly split 50/50 between personnel costs and operational costs. Operational costs include things like facilities and infrastructure, air operations, and ICT.

The facilities and infrastructure budget of $138 million for one year in Mali includes electrical transformers, water treatment and purification plants, wastewater treatment plants, solid waste management facilities, pest control and gardening techniques, roads, airports, and bridges.

The communications and information technology budget of $64 million for one year in Mali buys handhelds, mobile and base station radios, computers, printers, LAN/WAN networks, and geospatial data analysis. They also independently operate and maintain an independent network for voice, video and data communications, which includes 79 very small aperture terminals, 91 microwave links and 28 telephone exchanges.

Air operations budget for one year in Mali ($160 million) includes the operation and maintenance of a total of 74 aircraft, provision of 19,733 planned flight hours from commercial and military providers, and the training and enforcement of aviation safety standards. With the US$1.18

61 MINUSMA Fact Sheet, United Nations Peacekeeping

billion, tax payers in the US, China, Japan and elsewhere ensure that peace is kept for the people in Mali, while simultaneously investing in their local private defense industries and allowing them to innovate.

Next, we'll travel to Central Africa. MINUSCA, the peacekeeping operation in the Central African Republic was set up in 2014 and it currently deploys over 17,000 troops, military and police personnel. Much like in Mali, these people come from neighboring African countries, mostly Rwanda,[62] and use roughly half of the just over $1 billion budget.

The other half of the budget is spent on devices such as smart city cameras, sensors for surveillance and reconnaissance, quadcopters for high-resolution aerial surveys, small aperture terminals, computers, and printers. For connectivity, procurement includes LAN/WAN networks, voice over Internet protocol exchanges, point-to-multipoint microwave links and units, HF, UHF and VHF air band handheld portable radios and base station.

The Central African operation also uses licensed software such as topographic and thematic digital maps. Non-digital purchases are mostly cleantech products to ensure energy, water, and other infrastructure in difficult conditions.

Our third stop on this Africa journey is Somalia. UNSOS is technically no longer a UN peacekeeping operation, but a 'UN Special Operation.' After decades of war and other challenges, the UN has moved on to pioneering a new model for peacekeeping which they describe as "Support of UN and non-UN entities through use of 'light footprint' concept and outsourcing model." The light footprint model has proven to be a flexible and productive way of operating in high-risk environments, especially in maximizing the footprint of people on the ground. UNSOS currently supports over 37,000 troops with a support staff of less than 500.

What this means in UNSOS financial spending, much of the half a billion US dollars is spent on facilities and infrastructure ($94 million), air and ground transportation projects ($100 million), and ICT ($41.4 million).[63]

---

62 MINUSCA Fact Sheet, United Nations Peacekeeping

63 Approved resources for peacekeeping operations for the period from 1 July 2020 to 30 June 2021 : note / by the UN Secretary-General

If we make a bold assumption that all of the ICT purchases for these three peacekeeping operations drive innovation and research in the companies creating and producing these products, this alone would be a $600 million public innovation injection in just one year. I'm ready to make that assumption, as not all ICT is new innovation, but there is so much R&D going into cleantech, transportation, and human-related fields that would be excluded that it might just even it out.

You may argue that governments procuring devices from private defense companies is not innovation, it's just business as usual with existing products. My counterargument is that having one or more government contracts with several different governments allows companies to plan long-term. This means that they have a reliable stream of income, the financial leverage that this brings, and a customer who will never go bankrupt. I would call this a very strong government incentive for the private sector to innovate.

Sometimes things have surprising and unintended very long-term consequences. Here's a story from a Deep Pockets podcast episode on War and Innovation:[64]

> *On September 14, 1858, Mr. Fordyce Beals, an employee of Remington Arms Company, LLC received his U.S. Patent No. 21,478 for his design which required lowering of the loading lever to allow the cylinder pin to be pulled forward to free the cylinder.*
>
> *This ingenious patent was quickly implemented in Remington revolvers, owned by Philo Remington, which the company supplied to both the Union Army and Navy throughout the Civil War. More than 80 Union cavalry regiments were armed with hundreds of thousands of Remington revolvers.*
>
> *Around the same time a former Milwaukee newspaper editor Mr. Christopher Lathem Sholes experiments with a "writing machine". His writing machine is built from piano hammers, telegraph keys, and a pedal from a sewing machine. The writing machine was patented 10 years after the Remington rifle.*

64 Petra Soderling, How Wars fuel Innovation, Deep Pockets, November 25, 2021

By this time, the war is over and Mr. Philo Remington sits on a pile of cash. Mr. Sholes presents the writing machine to Mr. Remington, he gets excited and invests, and the Remington typewriter is born.

As the telegraph keys, piano hammers, and sewing machine lend themselves to the typewriter, the manual typewriter lends itself to the electric typewriter which lends itself to the word processor, again to the personal computer and finally to the smartphone.

So, in your opinion, who should we thank for the phone in our pocket: The Union Army, Mr. Sholes, or Mr. Remington? Like many things described in this book, this, too, is complicated. The purpose of this book is not to make complicated things seem easy, but to paint a fuller picture of how innovations are born.

Laura Thomas, a former CIA officer and Chief of Base, and a specialist in National Security Solutions, gives most credit to the inventor, Mr. Sholes. She also acknowledges that from a national security point of view, voters and taxpayers will want the inventors to have a support system. This is a completely new angle; private innovation is important not just for the economy, but for protection against adversaries. Ms. Thomas walked me through the early days of In-Q-Tel, CIA's own Venture Capital firm.

The Central Intelligence Agency CIA launched In-Q-Tel in 1999. I found a Wired article from 2003[65] that said that Silicon Valley was skeptical at first, but "now everybody's a believer." Over those 4 years a total of 3,400 companies had pitched their innovation ideas to In-Q-Tel. It has long been the country's only federally funded venture capital firm.

In Laura Thomas' words, an organization's decisions ultimately are a result of its culture and its people. There were some visionaries at CIA at the time who realized that this is exactly what the CIA needed, such as Susan M. Gordon[66] who was a driving factor behind In-Q-Tel. Susan Gordon worked for the CIA for 25 years and eventually became Principal

65 The Spies Who Fund Me, WIRED, September 1, 2003
66 Susan M. Gordon, Wikipedia

Deputy Director of National Intelligence. Ms. Gordon's role in the birth of In-Q-Tel supports our narrative that one person can make big changes, or set in motion a chain of events that will have a long lasting global impact.

Susan Gordon and others recognized early on the tectonic shift that technology was having and would continue to have on economic and national security, and the forces of innovation that were being unleashed in the late 1990s. At that time, post-Cold War, there was a real sense of promise and optimism in the US in general for new technologies and a feeling of being able to shape the future.

There was a recognition that the slow movement of the government acquisition process meant the CIA and the US Government were losing out on a lot of innovative technologies, and that the CIA could never afford to hire top Venture Capital talent. They would have to create a new way of doing things—and In-Q-Tel was born.

Today, In-Q-Tel serves not just the CIA but the entire Intelligence Community. They are looking into technology that has commercialization potential but also national security implications—AI/ML, quantum, biosynthetics, autonomous systems, data analytics, and cybersecurity are key areas. Famous In-Q-Tel investments and exits include companies like Palantir, D-Wave, Morpheus Space, Rigetti, and Xanadu. In-Q-Tel tends to invest longer-term than other VCs, not rushing to exit, and it has a 5% better performance rate than other lead investors (2018)[67].

The good news doesn't end there. In-Q-Tel doesn't only invest in private technology startups, it also sells these innovative teams and their IP to bigger companies. Google, Amazon, Oracle, IBM, Nokia, Samsung, Cisco, and Sony have all acquired In-Q-Tel funded technology and embraced it as their own.[68]

CIA is not the only government office that deals with national security and engages private companies in inventing new products and services. To maintain technological superiority on the battlefield, the Department of Defense (DoD) relies on scientific and technical knowledge developed

---

67 IN-Q-TEL, Unicorn Nest

68 Holden Page, What Big Tech Has Acquired From In-Q-Tel, The CIA's VC Arm, Crunchbase News, June 8, 2018

in large measure through research, development, test, and evaluation (RDT&E) funded by the department and performed by industry, universities, federal laboratories, and others.

For Fiscal Year 2022 DoD's RDT&E budget was $110.4 billion.[69] $2.7 billion (2.44%)[70] of this budget went to basic research covering microelectronics, quantum information science, advanced communications (mainly 5G), high-performance computing, missile early warning systems (the Next-Generation Overhead Persistent Infrared program), and Mobile nuclear reactor. All of this research is under the 'tech transfer' category, and will be done by private companies, universities and research labs in the United States and its ally countries and will directly support the economy.

The companies receiving this $2.7 billion from the DoD in 2022 owe gratitude to the Bayh–Dole Act[71] (the Patent and Trademark Law Amendments Act), which ruled for government entities to grant exclusive licenses to private companies for those inventions that remained in government ownership.

Other more recent US legislation that seeks to enhance national security and benefit the American private industry at the same time includes the CHIPS and Science Act, ORAN 5G Emergency Appropriations that secures domestic supply of telecom equipment, The Endless Frontier Act which seeks to maintain and build on U.S. science and technology leadership through investments in research and development, and Meeting the China Challenge Act of 2021 which helps businesses combat IP (Intellectual Property) theft, cybersecurity threats, and other China-related digital threats.[72]

How war and peace, or national security and peacekeeping if you prefer, fuels innovation can be a heavy topic. Social media has brought the war in Ukraine into our homes and lives in a way that TV was never able to do.

---

69 Press Release: Appropriations Committee Releases Fiscal Year 2022 Defense Funding Bill, The House Committee on Appropriations, June 29, 2021

70 Mitch Ambrose, FY22 Budget Outlook: Department of Defense, The American Institute of Physics, December 2, 2021

71 Bayh–Dole Act, Wikipedia

72 The United States Innovation and Competition Act of 2021, Senate Democratic Majority

Technologies invented or matured during World War 2.

Source: The National WWII Museum, New Orleans. Image credit Petra Soderling.

World War 2 created many innovations in use today

Many of us intimately follow the Ukrainians fighting for their lives, their families, their country and larger values. For months, the Ukrainian army had to fight with whatever technology they had available. Being inventive, they came up with their own solutions which the New York Times called the 'MacGyvered' weapons. These include retrofitted equipment such as missiles and rocket systems mounted on trucks and speedboats and plastic aircraft modified to drop grenades and other munitions.

> *"In the attack on the* (Russian missile cruiser) *Moskva, for example, the Ukrainians developed their own anti-ship missile, called the Neptune, which they based on the design of an old Soviet anti-ship missile, but with substantially improved range and electronics. They appear to have mounted the Neptune missiles onto one or more trucks, according to one senior American official, and moved them within range of the ship, which was around 75 miles from Odesa. The striking of the Moskva was, in essence, the Neptune's proof of concept; it was the first time the new Ukrainian weapon was used in an actual war, and it took down Russia's flagship in the Black Sea. (NYT Aug. 28, 2022)"*

Drones have been used in warfare for years, but they played a special role in Ukraine from the start. This has been the second war ever to use so called Kamikaze drones,[73] these are "suicide drones' that are equipped with explosives.[74] The Russian suicide drone is from the Kalashnikov Kyb family, unmanned aerial vehicles (UAV) that are four by one foot in size, carry a 6.6 pound explosive, and can fly for 30 minutes in one direction. They don't have any landing gear. This drone also does not have any sensors, including no GPS, so it is programmed just before taking flight.

Ukrainians are using Switchblades, a drone type made by American AeroVironment, Inc. The Switchblade is GPS-guided and an operator can control its flight sitting dozens of miles away. Like the Kalashnikov, the

73 Kyle Mizokami, Self-Destructing 'Kamikaze Drones' Are Hunting Down Targets in Ukraine, Popular Mechanics, March 28, 2022

74 Ameya Paleja, Russia and Ukraine to use suicide drones in armed conflict, Interesting Engineering, March 29, 2022

Switchblade can fly for 30 minutes; unlike the Russians, the Ukrainians can control and operate their drones during flight, find targets, where it crashes and explodes. The Ukrainian army is also deploying Turkish Bayraktar TB2 drones successfully, tweaking them to their needs. A 17-year-old Ukrainian Igor Klymenko won the Chegg.org Global Student Prize 2022 for developing a drone that detects landmines. His invention was selected from over 7,000 nominations from 150 countries.

For anyone watching or reading science fiction, the obvious next step in war technology is swarms of these kamikaze drones. As they are being developed with taxpayer money, the US government will make sure we get civilian uses for the swarms, too.

As is often the case when solving problems of national security, you don't get much information before the inventions are in use. I will just mention that the hours of research I put in this topic brought up research being conducted in smart materials such as two-layered graphene, Virtual Reality (VR) headsets and software, social media manipulation, quantum radars, drone missiles, robotics, microwave beams, and many others.

To be clear, I'm not suggesting war would be good for innovation. I am suggesting that the massive defense budgets that countries use are fueling innovation and benefiting the private sector during peacetime, and political decision makers have the power to control this money.

The economist Daniel Lacall wrote about this in his 2018 book 'Escape from the Central Bank Trap':

> *"The common perception that in some ways the war and the military industry were main contributors to innovation is simply exaggeration. Yes, for example the Internet and Silicon Valley were a direct result of World War II and the Cold War, but then so were the atom bomb and nuclear weapons ... This old-school approach to socio political capitalism worked for a time. It has not done so for more than two decades now."*

The fact that we are currently living through a war in Europe that is shifting defense strategies across the world is not a direct increased injection to innovation. Civil and military use inventions take place in government and privately funded labs all over the world. What has changed, perhaps, is the visibility of these technologies and a wide public acceptance to funding for peace.

KEY TAKEAWAYS:

1. Defense and peacekeeping budgets fund private innovation.
2. National security requires constant alertness and newest technologies to be developed and deployed.
3. Patient government risk investment can have a better performance rate than Venture Capitalists.

EDUCATION

I wanted to find out which are the most educated countries in the world. That turns out not to be a simple question with a simple answer. There are many different ways to rank education. Are we looking at a certain age population or everyone; should we include primary (K-8), secondary (high school, grades 9-12), and tertiary (college and post-graduate) education? Do we simply rank the percentage of educated population or should we also include the quality of the education? How attractive is this country for foreign students? Is the education system fair and accessible?

The World Population Review, an independent organization without political affiliations operating out of California, publishes nice tools for searching and ranking countries based on various education-related attributes. According to them:

> *"While education levels vary from country to country, there is a clear correlation between the quality of a country's educational system and its general economic status and overall well-being. In general, developing nations tend to offer their citizens a higher quality of education than the least developed nations do, and fully developed nations offer the best quality of education of all. Education is clearly a vital contributor to any country's overall health."*

The World Population Review presents us three different systems from separate sources to rank education in different countries. I will go through them in the order or scientific credibility.

First, the annual Best Countries Report[75] by the US News and World Report, New York based BAV Group, and the Wharton School of the University of Pennsylvania. This report interviews thousands of people across 78 countries on various issues. The education result compiles scores from three equally-weighted attributes: does the country have a well-developed public education system; would you consider attending university there; and does it provide top-quality education. In 2021, the top ten countries based on education rankings are the US, UK, Germany, Canada, France, Switzerland, Japan, Australia, Sweden, and the Netherlands.

For the US to come as number one raises eyebrows, and may be affected by the methodology of the study. The Best Countries Report is conducted by phone interviews of unsuspecting interviewees, which is not the most scientific method of getting data. Also, the fact that it puts equal weight on the question "would you consider studying there" as it does to the other two questions is likely to skew the international respondents select the US. Surely there are reasons for people to want to study in America other than its high quality educational system.

In actuality, among the 35 members of the Organization for Economic Cooperation and Development (OECD) the U.S. ranks 30th in math and 19th in science,[76] and unfortunately these results have been steadily dropping for the past decades. A drop in educational rankings goes hand-in-hand with the reduction of public spending in education. Budgets have not only decreased, but the smaller budgets are also distributed increasingly unequally within and among the 50 states.[77]

The second education ranking also comes from the US. This is the 'Countries with the Best Educational Systems - 2020 Global Citizens for Human Rights' list. Global Citizen is managed by Global Poverty

75 Gaby Galvin, In Denmark, Family Matters, U.S. News & World Report, November 21, 2022

76 Drew Desilver, U.S. students' academic achievement still lags that of their peers in many other countries, Pew Research Center, February 15, 2017

77 Aria Bendix, The US was once a leader for healthcare and education — now it ranks 27th in the world, Business Insider Nederland, September 27, 2018

Project, an international education and campaigning organization that is a registered 501(c)3 in the USA. They describe themselves as an action platform dedicated to achieving the end of extreme poverty, powered by a community of millions of global citizens.

This ranking emphasizes attributes such as low student-teacher ratios, kids staying in school longer, and graduating the greatest number of students with a quality education. The top ten countries with the best educational system are Denmark, Finland, Japan, Canada, Sweden, Germany, Israel, Netherlands, Singapore, and South Korea.

The third education ranking system is considered the most objective, and is the OECD Program for International Student Assessment (PISA). The last published results are from 2018 when some 600,000 students completed the assessment, representing about 32 million 15-year-olds in the schools of the 79 participating countries and economies. Instead of a questionnaire, these are actual computer-based tests that last for two hours and cover assignments for reading, mathematics and science. Anyone interested can go and take the test themselves at https://www.oecd.org/pisa/test/

The top ten countries for the combined math, science, and reading scores are, in this order: China, Singapore, Macao, Hong Kong, Estonia, Japan, South-Korea, Canada, Taiwan, and Finland. The United States is number 25. The top ten countries for just science scores are China, Singapore, Macao, Estonia, Japan, Finland, South-Korea, Canada, Hong Kong, and Taiwan. For science, the United States is number 18.[78]

The first thing you notice with PISA results is the high number of Asian countries. Estonia, Finland, and Canada stand out as the lone non-Asian countries on the top ten lists. The second interesting finding is that creating a good education system is not dependent on the size of your country. Singapore, Taiwan, Estonia and Finland are small countries that have consistently invested in their education; the results of this chosen strategy are clearly visible in their economic progress. The third finding is that publicly paid universal education is the best way to achieve generations of well-educated citizens, inventors, and employees.

---

78 PISA 2018 Worldwide Ranking – average score of mathematics, science and reading, Facts Maps

The most successful countries also invested the most taxpayer money. Macao and Finland both invest 6.3% of GDP, Canada and Estonia invest 5.3% of GDP and the rest of the top ten PISA countries all invest between 2.5% and 4.4%. In fact, a 2021 World Bank study supports the idea of investing in education. Based on data sets reaching back to 1970, The World Bank Comparable Returns to Education database found that in general, globally, there is a 10% return of investment in education. For men it is 9% and for women 12%. The return varies according to the level of education where primary education yields 11% return, secondary 7% and tertiary education yields a 15% rate of return.

Estonia is a real rock star among nations in many aspects, education being one of them. Estonia restored their independence in 1991, having been under Soviet occupation for decades. As a newly-independent country in the early 1990s, they rapidly and decisively started to build their country towards a western democracy and a market economy. In 1993, the per capita GDP adjusted to Purchase Power Parity was $11,909, in 2021 it had grown to $38,200. Today, the World Bank ranks Estonia as a high-income economy. Estonia joined the European Union and NATO in 2004, and in January 2011, Estonia adopted the euro and became the 17th Eurozone member state. Estonian leadership has always steered towards a balanced budget, and can now boast almost non-existent public debt, flat-rate income tax, free trade regime, competitive commercial banking sector, innovative e-Services and mobile-based public services.

Estonian children start at kindergarten from the age of three. The kindergarten is a playful experience where children are allowed to be children, with some formal playful learning introduced towards the age of six. The kindergarten prepares the children for school that starts at the age of seven, but focus here is on learning how to be social in a group, and how to get used to the daily rhythm of a school-like environment. The kindergarten teachers level up the attainment of children, and make sure they are emotionally and physically able to learn. As the whole age group goes through this same experience, there is a smaller gap in the PISA test results between rich and poor children that is seen in some other countries. There is no grading in kindergarten, but as they leave, the kids get a school readiness card that describes their skills and development.

At school, Estonia puts all levels of ability together, so the same age group that exited the kindergarten enters their first school year together. The universal basic education runs from age six to age 16. Throughout, every subject in school is taught in all-ability classes. In class, teachers have a relatively high level of freedom in designing their lessons and can take risks in exploring creative methods. There is a national curriculum that is set by the Ministry of Education and Research, but also flexibility for schools to adapt the curriculum to their local circumstances. Every child sits through every class, and within the first couple of years the teachers are expected to find ways of leveling up students from all backgrounds. When Estonia regained their independence in the early 1990s they looked at the Nordic model that was working well for children. This means a high level of equity. Every child is equal in the classroom and recess, as there is no cost for the parent for any of it. Textbooks are free, lunches are free, transportation is free, and education is free.

School is never detached from its surrounding culture, and this is also true in Estonia. Estonia is well-known for its digital advancements in both the private and public sector. The number of successful startups, digital voting, eCitizenship, filing taxes 100% digitally, and other exceptional progress has made Estonia the poster country not only for countries in Europe but beyond. At school, this demonstrates digital libraries, digitally-filed and -graded homework, and digital voting in school elections. Students are excited about technologies and entrepreneurship, and this shows an eagerness to do well in school, to learn and excel. The startup culture and fast-growing economy is an attractive ideal for young minds.

Invest in Estonia, the government's Investment Promotion Agency, says that 77% of Estonian students exhibit a growth mindset—they believe that they are capable of improving their intelligence and are willing to put effort into their own development in order to secure a better future.

70% of students plan to attain higher education with the most popular professions being ICT specialist, doctor, CEO, architect, and psychologist.

Getting back to the USA, how is it that the US so overwhelmingly produces the most innovative technology companies in the world when it's lagging behind other developed nations in education? There are several answers to this.

The USA is the home of some of the world's best privately-funded universities and research institutes. 64%[79] of students enrolling in tertiary education (that's any education after high school) pay tuition from private sources. Compare this to Finland (4%), Norway (6%), or Iceland (8%) to see that the system is different. Being a big country, these universities can skim the best students from across the fifty states, and provide excellent technical and commercial results. Students in these educational institutions don't all come from the USA, as we learned in the chapter on Immigration. By inviting already well-educated people from other countries to do postgraduate studies, or directly work for US companies is a big benefit. The US also fosters an entrepreneurial culture like nowhere else in the world. Silicon Valley, Boston, Chicago, Washington State, Boulder Colorado are all legendary tech hubs that other regions aspire to emulate. And for good reason.

Education can be a wonderful tool in a free, democratic world. It empowers people to take destiny in their own hands, and fulfill their wildest dreams. Education lifts families from poverty, and can create generations-long success stories. Well-educated citizens create new businesses, invent new products, and run for office in order to re-invest public assets to new generations. There are regions where those in power don't see that public education is necessary for everyone but prefer only that a select few receive a high-quality education. This is a short-sighted strategy, as we are clearly witnessing the virtuous cycle of high education and high financial wealth. Ensuring that entire generations in every corner of your country receive the best education you can possibly provide is unequivocally the easiest way for economic resilience and growth.

79 Education at a Glance 2021, Organisation for Economic Co-operation and Development, September 16, 2021

KEY TAKEAWAYS:

1. Investing in public education brings good educational results.
2. Good educational results correlate with innovation.
3. Innovation correlates with economic growth.

DIRECT FUNDING

There are big differences in how countries view direct funding given to private companies. By direct funding, I mean money being transferred from a government bank account into a company bank account, or money that the company no longer needs to pay thanks to government intervention. In very rough terms, you could identify two types of approaches: Large markets that support their existing businesses in a time of crisis, and smaller markets that incentivize startups to emerge and grow on a continuous basis. Let's take large markets first.

We spoke about the different actions governments took with Covid-19 in order to stimulate their economies in the chapter for Interest Rates and Taxes. In addition to these indirect tools, many countries also set up emergency funds that gave direct grants and loans to companies in distress.

China was the first country to experience the virus, and they quickly went into full alert issuing a number of measures to safeguard the health of the population as well as their economy. The South China Morning Post published an article on May 8, 2020[80] with the timeline and list of government actions starting from January 11, 2020. That's when the People's Bank of China cut banks' reserve ratio requirement (RRR) by 0.5 percentage points. The RRR, i.e. the money that banks are required to hold in reserve, was lowered to 12.5 per cent for big banks, and to 7

80 Frank Tang, What stimulus measures did China use to combat the economic impact of the coronavirus?, South China Morning Post, May 8, 2020

per cent for small banks. Starting from that day throughout 2020 there were weekly if not daily new aid decisions by the government.

China used all of the tools in the toolbox to keep the economy afloat, including direct subsidies, local government bonds to build housing and motorways (remember how we discussed these investment yields in the infrastructure chapter), tax exemptions and subsidies for consumers to purchase new energy vehicles like electric and hybrid cars. The city of Nanjing distributed purchase vouchers to consumers worth $42 million. Small businesses were allowed to postpone loan payments. The state ordered large banks to increase lending to small businesses by 30%. The list goes on.

Russia, another state governed by a strong central power, took a similar approach. In 2020, the country quickly deferred business loan paybacks and insurance payments, reduced the rate of mandatory insurance payments, and issued soft loans, direct grants and subsidies.[81]

Here in the US, decision making in the early days of Covid-19 was a tad slower (perhaps more democratic) but the same types of measures were eventually taken. The most widely used direct subsidy form was the Paycheck Protection Program (PPP) offered by the Small Business Administration (SBA). The purpose of this soft loan was to help businesses keep their workforce employed during the COVID-19 crisis. If you spent 60% or more of this loan toward payroll expenses and maintained the same level of employee and compensation level, the loan would be forgiven.[82] In total, PPP provided businesses with $800 billion in low-interest uncollateralized loans from April 3, 2020, through May 31, 2021. Almost all of these loans are expected to be forgiven. It is estimated the program preserved 2 to 3 million job-years of employment thus providing critical help to American businesses in a time of crisis.[83]

During "normal" non-pandemic times, the Small Business Administration runs other programs to help small businesses get off the ground and thrive. Its processes have been described as being the three C's; Capital, Counseling, and Contracts. SBA loans are issued through lenders, and

81 Governmental support to business in view of the coronavirus (COVID-19) influenced crisis in Russia in 2020, by type, Statista, April 2020

82 PPP loan forgiveness, The U.S. Small Business Administration (SBA)

83 Jim Probasco, Where the PPP Money Went, Investopedia, January 23, 2022

the role of the SBA is to provide government guarantees for the loans. In times of crisis, this guarantee can be up to 90 per cent. This is the 'Capital'.

As for the 'Contracts' piece, the SBA channels 23 percent of federal contracts to minority-, woman-, and veteran-owned businesses. These companies can stay in the program for nine years, during which the company can grow, hire, and make an impact in their local community. Each year about $50 billion[84] worth of government contracts are awarded through the SBA in diverse industries including defense, infrastructure, energy, IT, healthcare, education, and other fields.

There are about 900 SBA offices around the country, often co-located with colleges, business incubators and other entrepreneurial hubs. They 'Consult', coach, and run initiatives on the local level to help communities prosper. There's our third C.

Small Business Administration services

The SBA three C's	
Capital	**Up to 90% guarantee**
Counceling	**900 SBA offices**
Contracts	**$50 billion worth contracts**

Data source Small Business Administration. Image credit Petra Soderling.

Small Business Administration (SBA) serves American businesses in multiple ways

The 900 SBA business incubators are a good segue to the second category of direct government funding: To incentivize startups to emerge and grow. The Global Entrepreneurship Research Association, a private research think tank registered in the UK and Wales, collects and publishes an annual Global Entrepreneurship Monitor.[85] This extensive study tracks the vital signs of entrepreneurship around the world, and it includes

84 Small Disadvantaged Business, The U.S. Small Business Administration (SBA)
85 GEM Global Entrepreneurship Monitor

parameters for entrepreneurial framework conditions, many of which are controlled by the governments. It is an ideal tool for our purposes. These are nine parameters to track and rank government-influenced entrepreneurial framework:

1. What is the availability of and ease of access to entrepreneurial finance, including private equity and debt, and public grants and subsidies?
2. Are there high quality governmental entrepreneurial programs available?
3. Does government policy including taxation and bureaucracy support startups?
4. Do schools and colleges introduce entrepreneurship courses?
5. Can research be transferred into new business?
6. Is there sufficient and affordable physical, commercial and professional infrastructure?
7. Is the market free, open, and growing?
8. Do regulations encourage or restrict entry?
9. Does the culture encourage and celebrate entrepreneurship?

I'm mostly interested in the two first parameters, so I ran those searches. The top five countries for availability of public and private entrepreneurial finance are, in this order: Finland, Saudi Arabia, Netherlands, United Arab Emirates, and Sweden. The top five countries for high quality governmental entrepreneurial programs are, in this order: Saudi Arabia, United Arab Emirates, Netherlands, Norway, and Germany. The top ten best overall countries for entrepreneurs in 2021-22 are, in this order: United Arab Emirates, Netherlands, Finland, Saudi Arabia, Lithuania, Republic of Korea, Norway, Qatar, Switzerland, Spain, and the United States.[86]

If Finland is the best in the availability and easy access of entrepreneurial finance, including public grants and subsidies, let's see how they are doing it.

86 Chris Morris, Ranking The Best Countries for Entrepreneurs, Nasdaq, May 4, 2022

There are two main vehicles for public R&D funding. One is via the Ministry of Economic Affairs and Employment to support businesses; most of the funding here goes through Business Finland. The other is funding through the Ministry of Education and Culture to support scientific research.[87] Research funding is distributed directly to universities, as well as through the Academy of Finland. The Academy of Finland is the home of Finland's flagship research programs for 6G, Artificial Intelligence, biomaterials, precision cancer medicine, photonics, and social studies. They also run public research calls to fill gaps in national knowledge base.

The image below shows that both science and business have been receiving a steady €2 billion injection from the Finnish government each

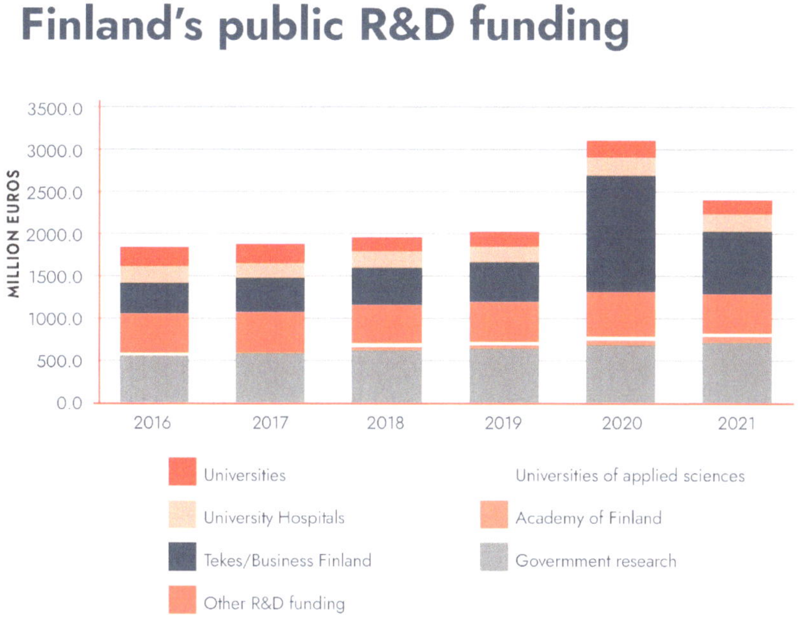

Data source Tilastokeskus. Image credit Petra Soderling.

The Finnish government grants R&D funding directly to private companies through Business Finland

87 R&D funding decreases in the 2021 budget compared to the exceptional year, Statistics Finland, February 25, 2021

year, up until Covid-19. In 2021, the Academy of Finland awarded €406 million for scientific research, and Business Finland granted €701 million to company-led R&D. This €701 million funded 3,912 R&D projects, 53% of which went to startups and small- and medium-sized companies (SMEs), and 47% to large companies. The role of large companies is important, as many of the co-innovation projects were led by one large anchor company who invited smaller, more agile and innovative startups to join. This type of funding creates a win-win for everyone involved.

Business Finland reports that those companies participating in its programs and receiving grants and loans increased their exports significantly, even during Covid-19. €800 million worth of exports from Finland to other countries in 2021 is calculated to have been initiated from Business Finland's support.[88]

Our second-best country for availability of entrepreneurial finance according to Global Entrepreneurship Monitor is Saudi Arabia. The Kingdom offers many incentives, and unlike the Finnish program, these are offered particularly to foreign nationals. For example, you are able to run your company in Saudi Arabia as 100%-owned by foreigners; there are no restrictions on foreign nationals to repatriate capital and profits out of the country; no sales tax; no value-added tax; no personal income tax; and no property tax. Your company is also able to carry losses forward indefinitely, and will be exempt from import tariffs for machinery, spare parts, and raw material. These are rare and generous perks that only a few countries will offer.

The government funded direct financing tool in Saudi Arabia is their Saudi Industrial Development Fund (SIDF).[89] SIDF was founded in 1974 to realize the objectives, policies, and programs of industrial development in Saudi Arabia. The 1970s was, as you may remember, a tumultuous decade for oil-producing countries, so this fund was aimed to restructure the country into a modern industrialized nation. The aim was and remains to support the private sector in various industrial fields, financing and developing the industry in conjunction with government entities. This wasn't entirely to found new companies, but also to scrutinize the existing

88 Results and Impacts 2021, Business Finland
89 The Saudi Industrial Development Fund (SIDF)

local industries in order to find the right government instruments for better performance. The first local businesses to enjoy these benefits were the electric companies, the date farming industry, hospitals, and other small- and medium-sized companies (SMEs).

Today, Saudi Arabia is delivering on its Vision 2030. As the program material states, Vision 2030 is "a roadmap drawn up by His Royal Highness the Crown Prince, to harness the strengths God bestowed upon us—our strategic position, investment power and place at the center of Arab and Islamic worlds." This massive program is built on six main objectives that stand on three pillars: 1. To build an ambitious nation, you need to enhance government effectiveness and enable social responsibility. 2. For a thriving economy, you need to grow and diversify, and increase employment. 3. Then, in order to build a vibrant society, strengthen the Islamic and national identity and offer a fulfilling and healthy life. There are 27 branch objectives supporting these six core targets, and 96 strategic objectives within the 27 branch objectives.

There is no overall budget communicated for the Saudi Vision 2030, instead the programs and pillars have their individual funding mechanisms. For example, The Public Investment Fund Program,[90] launched 2017 commits to injecting 150 billion riyals ($40 billion) into the economy annually until 2025. This program will then cumulatively contribute 1.2 trillion riyals ($320 billion)[91] to the non-oil GDP by the end of 2025. Another program relevant to our conversation, the National Industrial Development and Logistics Program was launched at the beginning of 2019 and commits to adding 1.2 billion riyals ($320 million) to the GDP by the end of 2025. Additionally, in July 2022, the Crown Prince Mohammed bin Salman announced[92] an adjacent target of annual spending in the research, development and innovation sector of 2.5 percent of the country's GDP by 2040. This new target will add nearly 60 billion riyals ($16 billion) to the GDP by the same target year.

90 National Industrial Development and Logistics Program, Vision 2030 Kingdom of Saudi Arabia

91 Alex Zhavoronkov, The New Saudi Arabia - Vision 2030 and AI, Forbes, July 14, 2022

92 Saudi crown prince announces 'national priorities' for research, development projects, ARAB News June 30, 2022

This is all well and good, but where is the direct funding for companies, you may ask. Saudi Arabia does not have a grant program, which the Global Entrepreneurship Monitor noted in their ranking review as a concern. Instead of grants or soft loans, the Saudi government offers moderately priced long-term loans. The SIDF loans are part of the National Industrial Development and Logistics Program. These loans are categorized by geographical zones, starting from a 15-year loan of 50% of a project, all the way up to 75% of a project for 20 years. Unlike Business Finland which only funds R&D, the SIDF funds operating costs. Just like the National Industrial Development and Logistics Program offers loans with favorable terms through the Saudi Industrial Development Fund, other programs, projects and objectives under the Vision 2030 have their own funding mechanisms.

These government initiatives are sending positive signals to the international and domestic private market. In the first half of 2022, investment in Saudi startups rose 244% to $584 million compared to the first half of 2021. There is a unique characteristic to the private investment market in the Kingdom—the informal investor. 16.1% of the adult population in Saudi Arabia considered themselves to be informal investors and this number is on the rise. The Saudi culture is very family-oriented which shows in the national investment charts; over 44% of private investments came from a close family member, 32% from some other relative, and the rest from colleagues or friends. Only 1% of the private investments received by entrepreneurs in 2019 came from a stranger.

This combination of visible government commitment to renewal, which attracts international entrepreneurs and affirms Saudi families to invest in local businesses, earned Saudi Arabia the second best place in the world for entrepreneur funding in 2021.

The Netherlands is the world's third best country for entrepreneurs to receive entrepreneurial finance. The World Bank keeps data[93] of government spending in R&D. Their numbers are based on UNESCO Institute for Statistics (I am mentioning this because I discovered that Eurostat

---

93 Research and development expenditure (% of GDP), UNESCO Institute for Statistics, The World Bank Group, June 2022

offers a different model for calculating the same percentage). In 2020, the Netherlands spent 2.29% of GDP in R&D, Finland 2.94% while Saudi Arabia spent 0.52%.

The Dutch government funds about a third of the total R&D expenditure in the country. In absolute numbers, they spent €7.3 billion in 2022, with 83% of the funds originating from two ministries; the Ministry of Higher Education, Culture and Science, and the Ministry of Economic Affairs. There are two ways of distributing these funds: Either as basic financing where the recipient has the freedom to choose where they use the money, or project financing given for a specific topic for a limited time and budget. The latter includes funding via the Dutch Research Council NWO, where researchers from a university or a company can apply for research grants.

The following image explains R&D expenditure in the Netherlands by performing sector, business, research, and higher education. The color

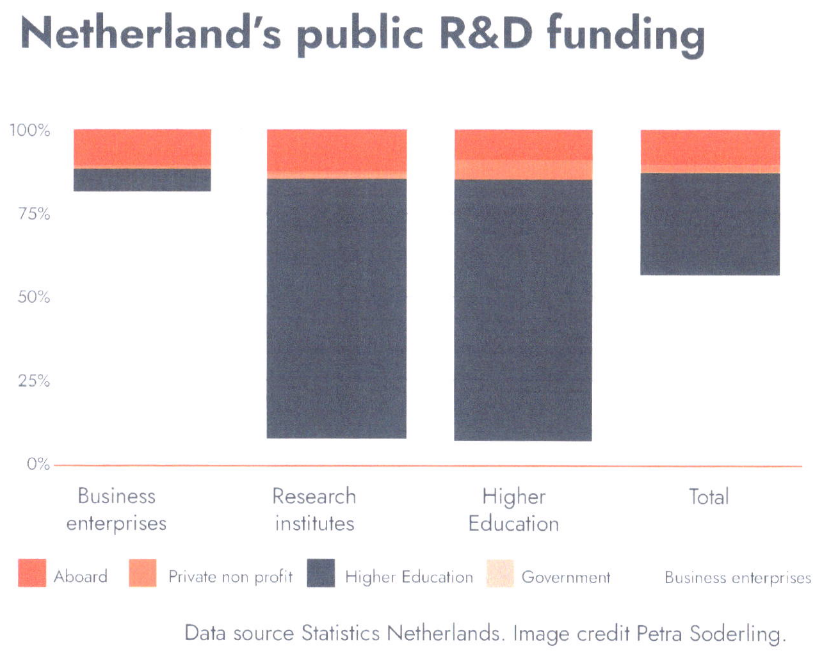

IMAGE 14 : R&D expenditure in the Netherlands 2019 by performing sector (x-axis), source of funds (y-axis), and percentage

codes are sources of funding for each sector, represented as their own columns. As you can see from the first column, the business sector funded most of its R&D itself, whereas research and higher education (second and third columns) are funded primarily by the government. The fourth column, total, shows how two thirds of R&D funds in the Netherlands come from private businesses, and one third from the government.

In euros, the total R&D expenditure in 2019, was €17.8 billion.[94] The business sector's share of this was €11.8 billion, the higher education sector spent €4.9 billion, and the research institutes €1 billion. An interesting finding in the Dutch model is that private companies also conduct much of the core research in the country. 44% of private company R&D is "experimental development", 45% is applied research, and 11% is fundamental research. According to the Rathenau Instituut definitions for science,[95] the 44% bucket is where the creation of new or demonstrably improved products, processes, marketing and organization happens.

For an individual entrepreneur, there are public-sector financing options available. That is the small sliver of government-originated funding in the first column of our image. These include incentives for startups, and subsidies or tax benefits for innovation and international trade. These options are available from local, regional, provincial, national and even European government bodies.[96] There are 23 different forms of R&D subsidies available for private companies, for topics as diverse as agricultural innovation, digital Europe, journalism, health, and sustainability. In addition, the Dutch government offers multiple targeted instruments for a startup to find support for the particular situation they are in.

Spending time on business.nl.gov you will quickly get the idea why the Netherlands is enjoying the status of third best place in the world for entrepreneurial funding.

The American, Finnish, Saudi Arabian and Dutch governments all support their founders, entrepreneurs, and companies to create new innovative products and stay afloat in times of distress. They do not take

94 Total R&D in the Netherlands by sector of performance and source of funds, R&D activity type and type of costs, The Rathenau Instituut, September 13, 2022

95 Definitions for Science in Figures, The Rathenau Instituut

96 Finding capital for your business, Netherlands Chamber of Commerce Financing Desk

equity in private companies; instead, they welcome the jobs, tax revenue, intellectual property and science.

Whether the government should ask for ownership in exchange for its support is an ongoing conversation. There are arguments for and against, and I have been on both sides of this discussion. In the final section of this book, we will weigh in on this question and help you form a strategy for your own country.

KEY TAKEAWAYS:

1. Direct funding can be a loan, grant, or a subsidy.
2. Government contracts for the private sector are forms of direct funding.
3. Most successful countries have programs to support private R&D.

INTRODUCTION TO SECTION 2

In the previous section, we learned how publicly-elected officials are responsible for passing laws and regulations that enable their local companies to grow and thrive while leveling the playing field for competition.

Nations can deploy interest rates, taxation and trade agreements to steer commercial behavior in a strategic direction. Those with some of the world's reserve currencies—the US dollar, the euro, the yen, the pound, the renminbi (RMB), the Canadian dollar, the Swiss franc, and the Australian dollar—have a slight advantage over other countries.

We also read about the role of immigration and education, two dynamics that determine how many smart and capable people you have in the talent pool. Some western countries with an aging population and low immigration are only now waking up to the harsh reality that you can only educate the children you have.

Infrastructure projects, including in defense-related industries, are an important enabler for existing businesses to be able to maintain operations and grow, and increasingly also to innovate new products and services.

Armed with these tools and the deep pockets they have, governments can ignite and expand innovation in their regions. But as Derek Thompson writes in his article "Why the age of American progress ended" (The

Atlantic, January/February 2023 issue), invention alone cannot change the world. What happens afterward matters.

Thompson writes that for many decades, the American government focused overwhelmingly on discovery rather than deployment.

> *After World War II, Vannevar Bush, the architect of our thrillingly successful wartime tech policy, published an influential report, "Science: The Endless Frontier," in which he counseled the federal government to grow its investment in basic research. And it did. Since the middle of the 20th century, America's inflation-adjusted spending on science and technology, through the National Institutes of Health and the National Science Foundation, has increased by a factor of 40.*
>
> *But the government hasn't matched that investment in the realm of implementation. This, too, was by design. Bush believed, with some reason, that politicians should not handpick nascent technologies to transform into new national industries. Better to advance the basic science and technology and let private companies—whose ears were closer to the ground—choose what to develop, and how.*

Thompson goes on to explain how we live in the world that Vannevar Bush built. The federal government invests billions into basic science and defense technology through initiatives like the National Science Foundation or the National Institute of Health, but for any civilian technology commercialization there has been a view that "Washington should fund the research and then get out of the way."

The issue with Vannevar Bush's strategy where government funds core research but does not support turning that science into sellable products, is that in essence the USA has become the R&D factory for other countries. It has not been able to turn its own scientific breakthroughs into commercial products that would benefit its own citizens. Solar energy, nuclear energy, microchips, radar, penicillin, GPS, and many others were American inventions made before the 1980s, paid for by the American tax payer, but they are being scaled up commercially by other countries. The difference between invention and implementation, argues Thompson, is that the former takes place behind the scenes but the latter is public. Therefore, tax payers and voters are rarely aware of the invention they paid for, but as consumers they buy and use the implementation of this invention every day.

As a lawmaker, you will need to be exceptional visionary to see how public investments into research create an economic return into the society. These innovation cycles can be long, much longer than one or two terms for democratically elected congressperson or member of a parliament.

In this next section, we are going to examine a handful of examples from five countries with five different technologies. These five countries are Canada, Brazil, Finland, Italy, South Korea. They are all nation states and in control of most of the tools we listed in section one. However, as we have established, the tools are not on-off surgical tools that can make precision cuts.

Investing in high quality education, as in Finland; relying on trusted trade partnerships like Italy did; allowing strategic companies to operate in a more flexible regulatory zone as in South Korea; building on decades of scientific knowledge as we will see with Canada; and licensing out government-owned intellectual property as Brazil is doing it—these are all long-term strategies. By implementing these and other similar strategies, these countries have been able to lift themselves out of poverty, rise in the charts as some of the most innovative nations in the world, and increase the prosperity and happiness of their people.

SECTION 2

EXAMPLES

WHAT THESE COUNTRIES HAVE DONE RIGHT

When I go to a conference I'm always looking for case studies and examples. I've been to numerous trade shows, scientific conferences and gala dinners. The thing that puts you to sleep immediately is a long, academic talk that's filled with industry jargon, acronyms, and does not immediately relate to the real world. Hence, I present to you, examples of what some countries have done right in their pursuit for more growth from more diverse sources, adding more science and technology based industries to their portfolio.

We've discussed the United States a lot and for good reason. It is the world leader in creating prosperity from technical innovations, prosperity that fuels a number of states as well as the federal economy benefiting every American. It is also a democracy, even if a flawed one[97].

In his book 'Principles for Dealing with the Changing World Order', investor Ray Dalio studies empires going back to the middle ages. He walks us through the Chinese, the Middle Eastern, the French, the Spanish, the British, the Japanese, the Mongolian, the Ottoman, the Dutch, the German, all the way to the American empire. He found symmetries with debts and

97 David Meyers, U.S. remains a 'flawed democracy' in annual rankings, The Fulcrum, February 14, 2022

interest rates in reserve currencies, and major political and social conflicts within these regions. Dalio argues that the repeating pattern of these phenomena is the causal effect for empires to behave in what he calls the 'big cycle'. These are waves in the world order that affect not just people living within the boundaries of said empire, but everyone dealing in economy, politics, religion, and science. The next big shift in the world order is China overtaking the United States in wealth, power and influence.

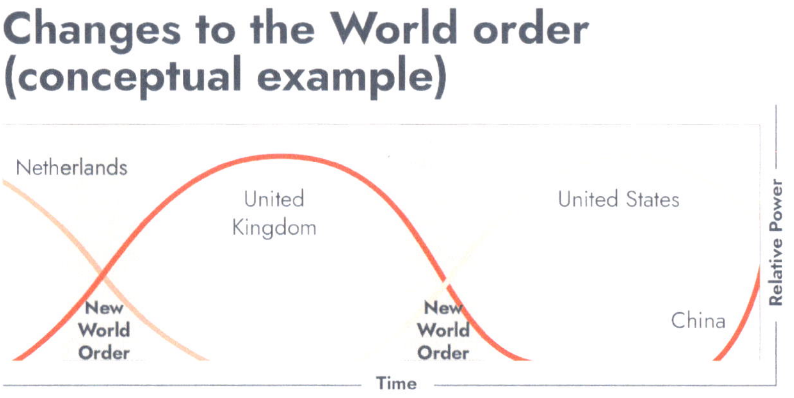

Image credit Ray Dalio, from Principles for Dealing with the Changing World Order.

Concept for changes to the World order by Ray Dalio

Barack Obama wrote in his memoir 'the Promised Land' that "If any country was likely to challenge U.S. preeminence on the world stage, it was China … And yet watching the Chinese delegation operate at the G20, I was convinced that any such challenge was still decades away - and that if and when it came, it would most likely happen as America's strategic mistakes." This was in 2009. I will leave it for political scientists and economists to argue whether China's imminent rise to power is due to long historical cycles, or America's strategic mistakes, or both.

Facts are, however, that in 2015 the Chinese Communist Party introduced a national plan they called China 2025. The Chinese Communist Party has been making five year economic plans since 1953, a model they copied from the Soviet Union, so having a plan was nothing new. What was new with this plan was the ambitious objectives that reached far beyond amounts of agricultural or manufactured goods produced, or number of new engineers educated. China 2025 was a plan to outgrow China's role as the manufacturer of the world's plastic toys, electronics, textiles, and raw materials and turn the country into a technology driven, high value add economy. Key industries to make this happen include IT, robotics, green energy, green vehicles, aerospace, high tech ships, railways, power and agricultural equipment, medicine and medical devices, and materials. As the name suggests this all was meant to be done by mainly Chinese raw materials with Chinese labor. At the time of writing this the plan seems to be going well. In 2015 The Global Innovation Index ranked China the 29th most innovative country in the word, in 2016 it had moved to #25, in 2017 to #22, in 2018 to #17, in 2019 and 2020 it was #14. In 2021 China was already the 12th most innovative country in the world, and the highest innovator in the S.C. upper middle-income group of countries.

As we learned in the section about Trade Agreements, the US and China have had a colorful history when it comes to tolls, tariffs, and trade agreements. During the Obama and Trump eras the US has accused China of benefitting from western technology even by stealing it. There's another side to the story that is less told. From 'The Battle over Patents - History and Politics of Innovation' by Stephen H. Haber & Naomi R. Lamoreaux:

> *What has garnered less attention from the interested publics, however, is that Chinese companies have also acquired foreign technology through copious patent licensing. Chinese companies operating in sectors such as transportation, energy, and robotics have paid top dollar to foreign patent holders to gain access to technology from the industrial frontier: Japanese and American firms have received billions of dollars in royalties in exchange for these licenses (Taplin 2018). China's royalty payments to the US grew dramatically faster than its GDP over the last two decades: IPR payments to US entities increased 25-fold while GDP … increased roughly 5-fold. In 2019 alone, China paid over $34 billion to the rest of the world for legal use of Intellectual Property (IP). The US accounted for 23 percent of this amount.*

In addition to relying on homegrown resources, licensing technology from the west, Chinese companies are also contributing to international standards and regulation bodies. Industry bodies such as IEEE, ITU, and 3GPP have large Chinese delegations[98] taking part in multiple working groups, work packages, BoFs, plenaries, etc resulting in Chinese owned technology as part of essential patents in critical technologies, or China preferred regulations over spectrum allocation, space law, or navigation. Any company anywhere in the world implementing these essential patents will have to pay a license fee to the Chinese IP owner. The Chinese plan is working. It is an economic success story we should pay attention to and learn from, but unless you are running a communist or authoritarian country you can directly implement only so many of their doctrines.

As we discuss different types of countries, their innovation strategies, investments and outcomes we need to be mindful of their regimes and resources. It is natural for larger countries to have high level, overarching strategies and plans with varying levels of government mandated, or government incentivized execution. When it comes to smaller countries, the successful ones often choose a category, or sets of categories, and systematically and methodologically build that capacity. Let's look at some of them.

In the next five chapters we are going to hear from five different countries, and how they made it. First, we will study how Brazil became an agricultural powerhouse, realizing they have an abundance of tropical climate but not enough scalable crops. By a clear and decisive strategy, the Brazilian government took steps that led to decades long success story in microbiology based agriculture.

Then, we will travel to Canada where the government realized their decades long investments into physics research is about to take off and created a national agenda to ensure the innovations will be implemented. Assisted with smart immigration policy, Canada is quickly becoming a world leader in quantum technologies.

In Europe, we will first go to Finland, and see how this small nation with not much other natural resources than forest demonstrates a return

98 Spectrum Wars - the rise of 5G and beyond, Jennifer A. Manner p. 48

on investment in their Artificial Intelligence strategy. Building on high quality education with a specific focus on engineering and computer science, direct government grants to businesses, Finnish taxpayers can now enjoy AI based revenue and services.

Our second European country, Italy, wanted to ensure their first early successes in the space technologies will become a nationwide source of pride. By working both with American NASA and the European Space Agency simultaneously, the Italian government ensured its teams would always be in the forefront of newest advancement, leading into unparallel commercial success.

The fifth innovative country in this section is the Republic of South Korea. South Korean companies have been pioneers in display technologies that we all use in our TVs, laptops and mobile phones every day. This is no co-incidence. Since the 1950s, the Korean government took multiple decisive steps to first get on par with the Western countries, and eventually in the early 2000s to surpass others in key digital technologies.

These five examples are meant to read as inspirational stories, not as direct how-to (that we will do in section 3). Sit back and enjoy the stories while couch surfing around the world in your mind.

KEY TAKEAWAYS:

1. There is a century old recurring movement on empires rising and falling
2. This movement is not divinely pre-defined, but man made
3. Regardless of the size and position of your country, you are benefitting of, suffering from, and influencing these outcomes

BRAZIL AND AGRICULTURE

In 2009, the leaders of Brazil, Russia, India, and China (BRIC) held a conference identifying that these four countries, the "Big Four", were on a path to becoming developed nations with a huge global economic impact. In 2010 BRIC became a formal institution, with South Africa joining the group, turning its name into BRICS. The one-person theory I keep bringing up explains that any big new announcement, any new product, company, law, or identifiable progress can always be traced back to one person who saw the future and set the ball rolling. With BRIC, it was Jim O'Neill.

In 2001, eight years before these countries' first-ever meeting, Jim O'Neill was an economist at Goldman Sachs. There he wrote a paper called "The World Needs Better Economic BRICs" being the first to coin the term. His argument was that these four countries hold over 25% of the world's land mass and 40% of its population. With such a powerful dual asset, they were a potentially dominant and leading force on the global stage. Leaders in these four countries read the article and called each other up to chat about it. The rest is history. Jim O'Neill later became the UK's Commercial Secretary to the Treasury so he was acknowledged in at least that way for his foresight.

I was doing my MBA between 2008-2012, reading up on global company strategies, international financing trends, accounting policies, B2B marketing, manufacturing, and innovation. The term BRIC has been engraved in the back of my head as something new, shiny, and promising from that era. Many things have happened since. These countries sometimes vote as a block in the United Nations, sometimes they don't. Economically and technologically, China has outperformed its BRIC colleagues as well as every other nation in the world. India is performing extremely well in lifting people out of poverty by educating a technical workforce and providing export-driven services, but not really delivering on the original BRIC promise. Russia is gradually slipping away from the world stage in matters other than military operations. No wonder we haven't read much about BRIC in the news lately.

One statistic from those times remains true: The land mass. Brazil has 65 million[99] ha (161 million acres) of cultivated land which is 8% of Brazil's land mass. To put this in context, another current world's breadbasket, Ukraine, has 41 million hectares of agricultural land which covers 70% of the country.

Before going into the breadbaskets of the world, let's go back in history for a moment. Brazil is the only Latin American country where Spanish is not the native language. That is, of course, thanks to the Portuguese being faster to colonialize Brazil than the Spanish were. The Portuguese arrived in the late 1400's and early 1500's and, as elsewhere, the coastal native tribes, Tupians, were killed by diseases brought by and bad treatment of the Europeans. There was no gold or other precious metal found on the coastland, but Europeans had discovered that by turning Brazilwood into powder you could die textiles a wonderfully luscious, deep red color. European aristocrats loved their red velvet.

In 1533, the Portuguese King John III divided the colony into fifteen separate blocks of land, each starting from the Atlantic coast with a border to its neighboring block in the north and south, but no western border. The king gave these land grants as gifts to favored people, the *donatários,* who could basically do whatever they wanted with the land. One of these land blocks included a populous area with vibrant villages and towns that is the present-day São Paulo. Another, Pernambuco in northeastern Brazil, was turned into a sugar-producing region. In addition to red velvet, the Europeans also really loved sugar.

99 Agriculture in Brazil, Wikipedia

Over the course of the 16th and 17th centuries, Portugal turned Brazil into an organized, productive colony by appointing royal governors, importing masses of slaves from Africa, and converting locals to Christianity ruled by the Jesuits. As the Jesuit missionaries, cattlemen, and explorers moved inland occupying more land, capturing native Indians as slaves, they also eventually discovered gold and other minerals. As decades and centuries went by, Brazil became a region rich with farmland, cattle ranches, and mining operations. The region unified by these economic advancements and a common language declared independence in 1822, called itself the Kingdom of Brazil, and put Pedro I as its emperor.

With the ensuing transition to a republic and a more stability in the young nation, the role of agriculture got even stronger. The sugar plantation owners in the north were forced to give up their influential power to coffee farmers in the south. During the 19th and 20th centuries, Brazil became known as the coffee provider of the world. Even as Brazil's economic structure has become similar to other Latin American countries with rubber, minerals, energy, and manufacturing producing much of the GDP, it remains one of the world's most important food producers.

Here's what The World Atlas 2017 writes as it ranks Brazil #1 in The Breadbaskets of the World:

> *Brazil has increasingly grown in importance as a breadbasket of the world, exporting five of the world's major crops. The country has two main agricultural regions. There is the Southern region which is more fertile with higher rainfall and the drier North Eastern region. The former produces a big percentage of the nation's grains, oil seeds and export crops while the latter produces cocoa and tropical fruits.*
>
> *Cattle keeping in Brazil is also a key agricultural enterprise, and the nation's annual beef production amounts to millions of tonnes yearly. The most important agricultural products in the country are coffee, wheat, soybeans, rice, sugarcane, corn, citrus, cocoa and beef. Brazil is the leading sugarcane producer in the world with a total yield of about 600 million tonnes every year.*

Of all the world's soybean producers, Brazil is ranked second. It is the leading exporter of meat from poultry and its pork exports are growing. The country also produces corn in plenty to feed its population and for export. Brazil is the third largest exporter of corn.[100]

Okay, so they have a lot of land and they use it for farming. Where's the innovation?

The innovation is at Embrapa, the Brazilian Agricultural Research Corporation. Embrapa is a state-owned research center that was set up in 1973 in order to develop farming technologies that work in the tropical climate. It develops technologies, knowledge, and technical-scientific information on farming and livestock. Embrapa operates as a for-profit company employing nearly 10,000 people, a quarter of whom are researchers. In 2017, they generated 3.36 billion Brazilian reals in revenue, which in today's currency would be over US$600 million. A 2007 New York Times article[101] called Embrapa "a world research leader in tropical agriculture" which is moving aggressively into areas like biotechnology and bio-energy. This is making Brazil a powerhouse of farming, an "agricultural superpower" as called by then Secretary of State Colin Powell. When the World Food Prize Foundation awarded two Embrapa-affiliated leaders in 2006, Nobel Peace Prize Laureate Dr. Norman E. Borlaug called their achievements "one of the great achievements of agricultural science in the 20th century."

So what are they working on? The problem that Brazil started to solve was this: They had masses land in the most fertile tropical climate, but there were no cultivated crops that would grow in such heat and humidity. So how to monetize this asset?

Their scientists in Manaus, a city in the middle of the Amazons are studying carbon sequestration, trying to understand if the so-called "Amazonian dark earth" soil types built up by pre-Columbian Indian tribes could be replicated in a lab. Unlike black soils in milder climates, tropical

100 Chelangat Faith, The Breadbaskets Of The World, The World Atlas, August 1 2017

101 Larry Rohter, Scientists Are Making Brazil's Savannah Bloom, The New York Times, October 2, 2007

soil often lacks the optimal amounts and combinations of carbon and phosphorus. Finding and patenting the right formula for tropical soil could be an enormous asset.

In addition to working on the soil, Embrapa has, to date, developed over 40 tropical varieties of soybeans, previously grown only in the world's cooler zones. This has led to Brazil being a top exporter of both pure soybeans, as well as high quality meat from cattle that has been fed these new soybean varieties.

In energy, Embrapa researchers are developing ways to, instead of pumping oil from the ground, growing diesel fuel. Sugarcane-based ethanol and palm oil-based fuels are well known, but Embrapa is working with 30 non-edible plants that could be used for similar energy purposes.

With livestock, they are working on hog types that survive well in the climate, are lower in fat and cholesterol, and produce a high yield of meat.

Embrapa scientists are discovering new uses for its rich tropical biosphere. In bioscience, imagine extracting natural polymers from spiders, as an example.

Embrapa's scientific work is divided into portfolios, programs, and projects. The current 34 portfolios are areas where scientists try to find solutions for national, institutional, and government demands. These areas range from industry (energy, fibers, milk) to solving horizontal national problems (drought, food safety, social issues) to technologies (nanotechnology, digitalization, genetics). Programs sit between portfolios and projects. The five program types include programs to support public policy, breeding and genetic improvement programs, institutional development programs, international programs, and co-funded programs. Projects, as the name suggests, is the organization closest to any individual researcher or other employee. Projects are entities with a start and end date, target setting, and allocated resources. There are thousands of projects that sit within the 34 portfolios and five program types.

Scientific core research is only half of the story, of course. The way to make science deliver its impact is through products, solutions, and applications—in other words, innovation. As a profit-making company, Embrapa offers its services directly to private companies for a fee. They

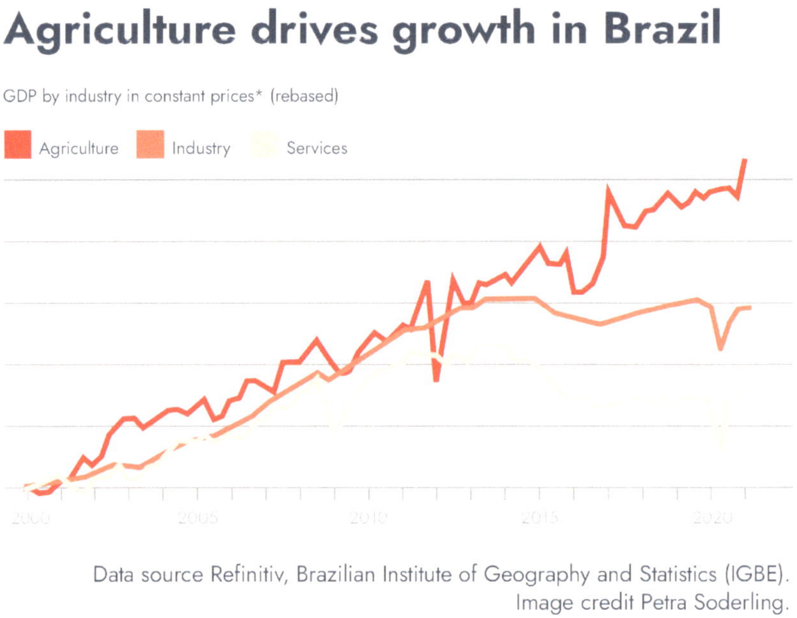

Science based agriculture has driven economic growth in Brazil.

provide lab analyses, product testing, monitoring, consulting and other services that help Brazilian and international companies perform better. They also accept joint R&D projects with the private sector, resulting in new deployable technologies, as well as license any intellectual property that Embrapa has created during any of these processes. Many of the licensing deals are made through public calls, in which licensing terms are defined for that particular technology in that call. At the time of writing (February 2023), there are 116 open calls to license Embrapa-generated assets. This technology transfer aspect of Embrapa—how its government-funded research is making a societal impact including generating commercial success in the private sector—is a unique success story.

Embrapa is a Brazilian Agricultural Research Corporation that reports to the Ministry of Agriculture and it is not the only piece this puzzle. A loose organization called the National Agricultural Research System (SNPA, the acronym from its Portuguese name), comprises of Embrapa, as well as other public institutions, universities, private companies, and foundations. This foundation has been critical to Brazil's economy. As can be seen from the Financial Times image below, although industry and services have been stagnant or even declining, agriculture has kept Brazil's economy afloat.

The World Bank issued a paper in 2014[102] that looked into the success of Brazilian agriculture. "Public Research Organizations and Agricultural Development in Brazil: How Did Embrapa Get It Right?" lists several factors for the success:

1. *Adequate levels of public funding. Embrapa's expenditures in the previous 20 years, at around 1% of Brazil's agricultural gross domestic product (GDP), compare well with figures of public spending on agricultural research and development (R&D) in more developed countries, such as Canada, the United States, and Australia*
2. *Sustained investment in human capital. 20% of Embrapa's budget was invested in the education and training of its employees between 1974 and 1982 alone. In 2014 three-fourths of Embrapa's 2,000 researchers held a PhD.*
3. *International collaboration and research excellence. From the beginning, researchers were drawn from leading universities, setting a high standard of research excellence. Furthermore, Embrapa strengthened its international links by establishing "virtual labs abroad" on three continents to institutionalize knowledge generation and exchange.*
4. *A mission orientation and IPR policy. Embrapa was created with "the mission to provide feasible solutions for the development of Brazilian agribusiness through knowledge and technology generation and transfer." Curiosity-driven research was discouraged by using agricultural research as a means to solve concrete problems faced by the Brazilian farming sector. Pursuing an open innovation system and IPR policy in the agricultural*

102 Paulo Correa and Cristiane Schmidt, Public Research Organizations and Agricultural Development in Brazil: How Did Embrapa Get It Right?, The World Bank, June 2014

> *sector facilitated technology transfer, diffusion of new cultivars, and the filing of international patents. An IPR policy that favored social well-being rather than benefiting just corporations allowed new technology to be distributed at production costs only.*

While the World Bank reserves the notion that directly copying this model to another country may not work in a similar manner, they go on to write that "*These factors may be used as guiding principles for the design of public organizations providing agricultural research*

and extension services in developing countries." I think this is as close to a good endorsement as any national strategy-driven research organization could hope for.

As we are inspired by Embrapa's success, let's remind ourselves why we are discussing these themes. For any country, state, region or city, you need a public strategy for how you are going to ensure future success for your voters and taxpayers. As you look at the current situation and assess the best way forward, think about the resources you have: your natural environment and your people.

Even when you don't have all the pieces of the puzzle, identify the ones you have and the ones you lack, and start building success from there. Allocate patient funding, invest in education, reach out for international collaboration, and be smart with the intellectual property your investment buys. With these encouraging examples, you may be able to build an industry that will carry your country forward through tough times, just as Brazil's agriculture sector has done.

As for Jim O'Neill, here's what he had to say 20 years after writing the inaugural BRIC paper: "*Two decades on, the BRICs promise lingers.*"[103] China is moving on its own superstar path and outperforming its BRIC family members. India, according to O'Neill, is broadly developing along the path of the original BRIC. In 2021, (post-Covid but pre-Ukrainian war) O'Neill placed Brazil and Russia into one group that "*Brazil and Russia have both suffered from the well-known commodity curse and, as*

103 Is the Emerging World Still Emerging, Jim O'Neill, International Monetary Fund, summer 2021

evidence suggests, are far too dependent on the world commodity cycle for their own sustainable development." The answer for these emerging markets to keep emerging into sustainable development is diversification. For Brazil, what better way to diversify than to innovate for new products in new industries from its natural resources—their land?

KEY TAKEAWAYS:

1. Even a developing country can become an economic superpower in one sector through R&D and science.
2. Success comes from relying on your strengths and solving local problems.
3. An adequate sustainable amount of public funding and education are keys.

CANADA AND QUANTUM TECHNOLOGIES

I discovered quantum computing in 2020 when my native Finland announced they would build a 20-qubit quantum computer. What in the world is a quantum computer—and what is a qubit?

By then, I knew that *quantum* had something to do with a cat in a box, Harry Potter's invisibility cloak, and possibly the Higgs boson aka the God particle. I had to look into this. Before I knew it, I became fluent in terms like entanglement, superposition, lattice, superconducting, ion trapping, cold atoms, fault-tolerant, error correction, quantum supremacy, and post-quantum crypto (just to show off with a few). At that time, my job was to attract investments from North America to Finland, so I had to learn the ins and outs of the quantum industry so I could have credible conversations with American quantum companies. I fell in love.

So, what is it then? The best way to understand how quantum will change everything is through use cases. First, there's the computing part. Quantum computers will be so much more efficient than classical computers that they can solve problems that are currently practically unsolvable. The Traveling Salesperson problem, for example, was formulated in 1930 and is one of the most studied mathematical optimization problems to date. It goes like this: "Given a list of cities and the distances between each pair of cities, what is the shortest possible route that visits each city exactly once and returns to the origin city?" Depending on the number of cities, or any nodes in any network, today's classical computer, or even today's supercomputer will need from years and decades to hundreds of years to solve it, whereas a quantum computer will do it in seconds or a fraction of a second.

Now juxtapose this computing power and speed with drug development, financial markets, energy, space exploration, and telecommunications. You will start to realize quantum's disruptive power.

Another eye-opening example is in quantum sensing. The very essence of quantum mechanics is a phenomenon called superposition. Superposition is a state where quantum systems such as entangled atoms can exist in several separate quantum states at the same time. Sometimes you see this referred to as being one, zero, or one and zero at the same time. Classical bits that we use today can only be a zero or a one but never both at the same time. The thought experiment with Schrödinger's cat[104] illustrates this paradox: A cat is put in a box with poison. Before you open the box and discover the cat to be either dead or alive, the cat is both dead and alive at the same time. With entangled atoms in a superposition, it is theoretically possible to construct an invisible sensing bridge between two quantum systems that are in the same state regardless of their physical distance. Think of one walkie-talkie on Mars and the other on Earth as being, in fact, the same walkie-talkie containing the same set of information simultaneously. Research on migratory birds and whales is discovering that their navigation systems are a complex web of quantum sensing and the earth's magnetic field.[105]

Quantum mechanics is opening the door to not just understanding but replicating "the sixth sense". Blown away? We should be.

Brian Lenahan, the founder of Quantum Strategy Institute, is one of the visionaries advocating for practical everyday uses of quantum technologies. He talks about how businesses should look at this promise of faster, more secure, and more accurate computing, and then make their

104 Schrödinger's cat, Wikipedia

105 The sixth sense: How animals use the Earth's magnetic field for orientation, University of Oldenburg, July 14, 2022

roadmaps, prioritize major use cases, and think about resourcing. When you listen to him speak in his soothing voice in a hopeful, optimistic tone you cannot help but believe. Brian is one of those people who are both convincing and inspiring at the same time because they themselves are convinced and inspired.

This sentiment, this dedication, comes from experience. For 25 years, Brian Lenahan worked for a major bank, living through the 2001 dot-com bubble, the 2008 recession, and other, smaller crises. He told me that during those 25 years, he held 14 different jobs whose common denominator was always solving problems for clients. Get a new role, solve a problem, and move on. Get a new role, solve a problem, and move on. Rinse and repeat. That's how you become a person who looks at life through a lens of serenity and confidence. If there's a problem, we can solve it. If there's an opportunity, we should look at seizing it. Brian is probably raising his eyebrows at my over-excited introduction of the sci-fi-sounding possibilities of quantum, as he is not one to hype. "It's not about the billions invested or the much-hyped breakthroughs. It's about the consumers creating business,", he said.

I'm still going to talk about the billions, though. In the US, the National Quantum Initiative Act was passed in 2018 and earmarked $1.2 billion for R&D. France announced its quantum strategy in 2021 with a €1.8 billion budget. Germany's plan the same year received a €2 billion funding allocation, while Japan's quantum strategy in 2020-22 came with a target of 10 million Japanese quantum users by 2030.

According to KPMG in January 2022, there were 15 countries with a quantum strategy, coupled with US$23 billion allocated for R&D.[106] Quantum is red hot with governments, and for good reason. While quantum physics has been a subject in universities for decades, it is only recently that scientists have been able to get close to commercial use of quantum technologies for computing, sensing, cryptography, and other uses of Quantum Information Science, QIS. There are two main drivers in this race among countries: Grab a leadership position in this new technology area; secure national and private assets against a quantum attack.

106 Unlocking the Business Value of Quantum Computing, KPMG, January 17, 2022

We've already discussed Canada's immigration policy, but how did this modern, open-minded, forward-looking country become what it is today?

Unlike in Brazil, the first Europeans to arrive in this part of the world were not European colonists: Vikings came first. The Vikings who occupied Greenland 1,000 years ago sailed west and spent time in Labrador and Newfoundland. The French arrived in the 16th and 17th centuries only to battle harsh climate conditions and native Indians. The relationship with the Iroquoians was especially difficult, with the French aligning themselves with the Algonquin, Montagnais, and Huron tribes against the Iroquoians. After the French and the Iroquoians made peace in 1701, they began to collaborate in the fur trade and built a thriving colony of French-speaking Roman Catholics.

As we know the battle over who ruled North America was fierce between England, France, and Spain, with borders shifting over centuries. When the Brits took over Quebec from the French in 1759, they allowed the French-speaking "Canadiens" (from the Iroquois word Kanata meaning *village*) to continue their French lifestyle. A few years later, in 1776, when thirteen colonies south of the "Canadiens" declared independence from the British king, tens of thousands of Dutch, German, British, Scandinavians, and even aboriginal Mohawks who were loyal to the British king, moved north. This group included some 3,000 black freedmen and slaves. Over the 18th and 19th centuries, Canadians established and grew international trade around fur, fish, and timber.

One of Brian Lenahan's ancestors came over in the 1820s with the British army and landed in Nova Scotia on the Atlantic coast. Other ancestors of Brian's arrived in Canada from Ireland in the 1840s. As waves of immigrants continued to flock to North America, there was a massive move from east to west. Like many others, Lenahan's ancestors moved to current-day Ontario in the 1840s. This region was then called Canada West and doubled its population over the next decade, making the English-speaking population larger than the French-speaking one. An economic boom followed, railways were built, and Brian's family lived comfortably as cattle farmers and dairy farmers. The brave people of Ontario, Quebec, New Brunswick, and Nova Scotia received independence from the king as the Dominion of Canada when the British Parliament passed the British North America Act in 1867. Brian's family counted among the many who celebrated on that July 1st.

Today, Canada is the world's eighth-largest economy. Even though its natural resources (minerals, timber, oil, and gas) play a significant role in creating jobs and offering a nice standard of living, services are increasingly important. Three-quarters of Canadians work in the service sector, including 1.2 million high-tech workers.

The Canadian government has had a clear vision of the tactical steps to take to make Canada a pioneer in new technologies, to use technology to benefit Canadians economically and socially, and to build a private sector that is future-proof in all respects. In 2019, Canada invested CAD$40.3 billion in domestic research and development, of which CAD$7 billion was provided by the federal and provincial governments. These scientific investments have resulted in a total of fifteen Nobel laureates in physics, chemistry, and medicine.

Quantum fits this story perfectly.[107] Between 2009 and 2019 the Canadian government invested over CAD$1 billion (US$750 million) into quantum research. By 2017, these investments had pushed Canada to the global 5th place in total annual expenditures on quantum science, and 1st place among G7 nations in per-capita spending on quantum research. In 2016, Canada was 5th among G20 nations for patent filings in the area of quantum computing and telecommunications.[108]

All these scientific publications, theories, and patents are not enough, though. As Brian Lenahan points out, the transition from public investment to improved quality of life for citizens isn't complete without commercial success. It is businesses that take an input, a patent, or another scientific finding and figure out how to multiply its value. Canadians have certainly understood this, watching capitalism boom and prosper across their southern border, but it's only recently that the government

107 Lori Valigra, Canada lays the groundwork to become a powerhouse in quantum technology, Science|Business, June 23, 2022

108 Ben Sussman, Quantum Canada, Quantum Science and Technology, February 22, 2019

has supported the commercialization of science. The new rounds of public investments into quantum technologies are putting a bigger emphasis on productization.

Quantum physics, quantum mechanics, and quantum information are not fields you can just jump-start from zero with a government mandate. In order to create meaningful scientific or commercial impact, you will need labs, equipment, know-how, and talent. All fifteen of the countries in the KPMG list have a long history in physics research, and most have a long history in quantum physics, nanotechnologies, computer science, photonics, microelectronics, and other adjacent fields. So does Canada.

From a 2019 study by IOP Science[109]:

> *Canadian scientists have been pushing the boundaries of quantum research for many decades. In 1984, Gilles Brassard at the Université de Montréal and Charles Bennett at IBM Research introduced the first scheme to encrypt information using the properties of quantum mechanics. Brassard is rightly known as Canada's 'father of quantum information' because of this pioneering work. Since then, Canadian scientists and research institutions have further advanced the science through discoveries in areas that include the following: photonic quantum computing, measurement-based quantum computing, mathematical aspects of quantum information processing, solid-state quantum computing, adiabatic quantum computing, quantum metrology, quantum communication, quantum networks, quantum cryptography and related software, as well as quantum materials and quantum-material based devices as a foundational enabling technology. These advancements have been accompanied by progress in development and refinement of several physical platforms for quantum technologies, including photon-matter interfaces, spin-based qubits, and superconducting qubits.*

As you can see, there is an extraordinary history of science there, so it was only natural the government wanted to enhance this critical field

109 Ben Sussman, Quantum Canada, Quantum Science and Technology, February 22, 2019

within the nation. In 2021, the Ministry of Industry commissioned a study from ISED (Innovation, Science and Economic Development Canada), which was published as "National Quantum Strategy Consultations: What We Heard Report".[110] This research was done in concert with the CAD$360 million budget for quantum in that year. The consultations were held as roundtable discussions with over 70 stakeholders from small and large private businesses, academia, the government, and non-governmental organizations. In addition, 240 individual citizens supplied input via an online questionnaire. The outcome of these consultations resulted in a series of findings, or silent joint agreements if you will:

- *The time to act is now, across all fronts, including research, talent, commercialization, international and security.*
- *Talent is critical, as we are facing strong international competition to recruit and retain talent.*
- *It is premature to pick winners at this stage in quantum's development, but large investments are required.*
- *The overall amount of NQS (*National Quantum Strategy*) funding may be insufficient to achieve our goals, especially as other countries have promised to invest more.*
- *Improved communications (including a branding strategy) are important to convey the advantages of studying, doing research and operating a business in Canada.*
- *Engagement with end-users and consumers is needed to build awareness and address concerns around adoption of quantum technologies.*
- *Domestic and international collaboration are essential for Canada to remain a leader in quantum.*
- *Canada's domestic market is small, so we need to connect with researchers and businesses in other countries right from the start.*

Having read a number of government strategies and action plans (and contributed to a few), the fifth bullet stands out. Canada wants to

110 National Quantum Strategy Consultations: What We Heard Report, Government of Canada

improve its already successful communications about the advantages to study, do research, and run a business there. This is fully in line with their immigration policy, which, as previously discussed, is unique and highly successful. Complemented with the two last bullets, international collaboration and access to other markets, this reads as a very open and collaborative effort to build something big, something important, and to do it together.

In my interview with Brian Lenahan, I asked him what it was like growing up in this northern nature paradise, and what shaped him into eventually becoming a spokesperson for commercial adoption of quantum and AI. His father tried to make Brian a farmer to carry on the family tradition—many times in fact, I was told—but his curiosity towards the world wouldn't allow it. Brian told me that since the first levels of school science, books, and the general smorgasbord of knowledge and information made him hungry for more. Perhaps because of this highly sophisticated education system or maybe because of the vast wilderness of the Canadian landscape, young Brian also felt an exhilarating freedom to be able to do whatever he wanted.

As he completed his studies and took on new jobs and new adventures, Brian moved from east to west, from Ottawa to Kingston to Toronto to Mississauga—just like his immigrant ancestors had done 200 years earlier. It is no wonder the Canadian story resonates with smart people around the world looking to move to a country where they can better themselves and their families. The story is genuine.

All of this is well-reflected in the National Quantum Strategy Consultations: What We Heard Report. It discussed the previously-defined National Quantum Strategy (NSQ) pillars and themes on the importance of aligning efforts and ensuring equitable access to quantum technologies, commercial scale-up of Canadian quantum companies, diversity and inclusion in the quantum workforce, and securing supply chains. There were also concrete recommendations on the establishment of an advisory body, how to address intellectual property issues, and participating in international standardization efforts.

All of this effort, of course, intended to bring prosperity and a good life for Canadians. In a 2021 study, the National Research Council of Canada estimated that the quantum sector would become a CAD$139 billion industry in Canada with more than 200,000 jobs and CAD$42 billion in returns by 2045, potentially contributing 3% to Canada's GDP.

But in the deep, hard technology fields like quantum, the return on investment cannot be calculated purely by the number of jobs or direct economic impact. By betting on an entirely new technology which is potentially one of the most disruptive technologies of our lifetimes, a country can create a position that will fuel indirect benefits that we cannot even imagine yet.

Let's consider times past when the USA invented the transistor or the internet. No one was able to imagine the commercial applications that would come years later and thus, there was no way to calculate the return. There were, however, visionaries who thought it was important for the American government to invest in these technologies, to ensure they are in line with national security interests, and to keep fueling the success of American private industry. Google, Facebook, Microsoft, Amazon, and many other American private success stories happened because of those visionaries.

Quantum is the transistor of the 1950s and the internet of the 1990s. We know it will change everything; we just don't know exactly how and when. Canada has decided to sit in the driver's seat and own this space from science to innovation to monetization. Brian Lenahan agrees with the strategy: "The 'when' in quantum is not a date, it's a range. It is already generating revenue in some businesses and will only grow from here." His goal is to continue to accelerate the growth of quantum by providing balanced and independent views through the Quantum Strategy Institute and his books.

As for other incredible future technologies, Brian's list is long. DNA storage, human brain-powered computers, 6G-enabled virtual worlds, space travel, biomaterials, 3D printed or lab-grown organs, and so on. "There will always be new books to write."

KEY TAKEAWAYS:

1. Quantum technologies are at the forefront of technology disruption.
2. Canada has strategically laid out its quantum path.
3. Investments into disruptive deep technologies will have a spillover effect to other fields.

FINLAND AND ARTIFICIAL INTELLIGENCE

I hold two passports, one Finnish and one American. One of the triggers for me to write this book was all the learning I accumulated while I worked for the Finnish government—how important public R&D funding is and the many ways government helps startups cross the valley of death.

This chapter is also a bit embarrassing for me to write since I did found a startup in Finland and it did receive government grants. What I did then was to move to New York City and take my business with me to the United States; I never returned to my native country. Maybe saying nice things about the Finnish model is my way of giving back.

So let's get to it. Finland's impressive modern digital economy builds on decades of excellence in electronics. How did this happen?

Before World War II, the country was a small northern European agrarian culture with GDP per capita less than half of the UK or the US. 70% of people worked in agriculture or forestry in 1900.[111] The main exports up until the 1950s were metal, wood and paper, chemical, and food products. Low-tech stuff. Apart from wood, Finland doesn't have rich natural resources such as coal or oil, and it has relatively few minerals. Although it has a lot of water in hundreds of thousands of lakes,

111 Riitta Hjerppe, An Economic History of Finland, University of Helsinki, February 10, 2008

it is a flat country with little possibility to create hydro power. Its largest copper mines were depleted by the 1980s. Immigration into the country has been low. How depressing a picture is that?

However, as with most European countries, everything started to look different after World War II. Finland had lost territory to the Soviet Union, and some 400,000 refugees needed to be re-settled into the remaining land. Housing needed to be built; land needed to be cultivated. The country took its reparation debt to the Soviets very seriously and quickly set up machinery and shipbuilding industries. Also, wood exports to other European countries rebuilding themselves quickly soared.

Internationally, Finland joined the World Bank, IMF, Bretton-Woods agreement, the General Agreement on Tariffs and Trade (GATT), and the European Free Trade Area (EFTA). These memberships created commercial successes with the west, while the country balanced its large import and export trade with neighboring Soviet Union.

One of the most devastating Soviet bombings in Helsinki during the Winter War had hit the Technical University on November 30, 1939. At that time, the Technical University was an industry-funded institution, one of many that helped the Finnish armed forces come up with inventions. It was sort of like an industry-driven startup incubator but for a patriotic cause. The university also helped the home front by testing food safety and creating substitute products during the shortage period. While the bombing of the university building slowed down their research operations for a while, the motivation for the teachers, students and scientists to get back to work only grew stronger.

President Risto Ryti and his administration realized that for Finland to prosper when the war ended, they needed to safeguard and nurture this nascent technology field. In 1942, the government launched a public, nationwide research agenda that included the founding of Valtion Teknillinen Tutkimuslaitos (VTT), the Technical Research Centre of Finland. The Technical University also shifted to become a publicly-funded institution offering higher education in engineering, architecture, chemistry, material sciences, physics, natural sciences, and—most importantly to our story—electronics, communications and automation.

The electronics industry grew in size, exports, and job creation throughout the second half of the 20th century. The catalyst for this was the President's vision back in 1942. In the early days, electronics was seen as a supporting function to the forestry industry. Decision makers thought that in order to succeed in the highly competitive global wood and paper market, Finland needed to find a competitive edge. With the government backing them up with infrastructure and education, several private companies started inventing, patenting, and creating new types of electronic products.

The company Strömberg (later part of ABB) developed electronic control systems for the forest industry and for logistics. Logistics itself was important to get paper products distributed from remote Finland to consumers in Central Europe, Soviet Union, and North America. The company Kajaani-Elektroniikka (later part of Valmet Automation) developed measurement and analytics technology for the manufacturing of paper pulp. Valmet Instruments (later Valmet Automation and Metso Automation) created electronic automation systems and measurement instruments for the timber-related processing industry. Even Nokia, perhaps the best-known Finnish company even today, started its electronics unit to serve its pulp and paper unit.[112]

What ties these stories together is that all of these projects became successful electronics products on their own, creating a new industry that was completely separate from wood and paper.

Strömberg was an important company to Finland in multiple ways. Its control system development led to an invention that became a cornerstone of Finnish electronics—the frequency converter. The frequency converter allowed the manipulation of speed in electrical engines which radically changed the operating principles in vehicles and manufacturing. In the early 1970s, Strömberg also created the first industrial computer that was used for process control purposes. The 'Strömberg 1000'[113] (also known as the 'SELCO 1000') ran at least three programming languages—Assembler, Monet 22, and Algol 60.

112 Little Finland's Transformation to a Wireless Giant, Petri Rouvinen and Pekka Ylä-Anttila, ETLA, The Research Institute of the Finnish Economy

113 Strömberg 1000, Wikipedia

The seed was planted. Hunger for new inventions, more R&D, and newer ways of working was a driving force in the 1970s and 1980 that led to explosive, even exponential growth in electronics (clearly visible in the image below). This growth especially since the early 1990s was predominantly created by one company called Nokia. In 2000, Nokia accounted for four percent of Finland's GDP, 70% of the Helsinki stock exchange, 43% of the country's corporate R&D, 21% of exports, and 14% of corporate tax revenue![114]

Many books have been written about the rise and fall of Nokia. This is not that book. This book is telling the story of the public decision maker and what they can do to boost their economies. In war-torn Finland in the 1940s, the government decided to invest in free primary, secondary, and tertiary education, in free universal healthcare, and in infrastructure. The government negotiated international deals for commerce, funding, monetary systems, tolls and tariffs. The administration supported both industry and workers in order to create not just an economy, but a society where everyone willingly works together. All of these decisions combined created the platform used by Nokia to rise, thrive, and eventually enable Finland's current digital innovation economy.

Speaking of digital innovation economy, this would be a good time to introduce our Finnish lead character who kindly shares with us her vision for Artificial Intelligence. We'll start from the 1990s.

IRC, Internet Relay Chat, is a text-based instant messaging developed in Finland in 1988. This pre-internet and pre-cell phone era chat was wildly popular among Finnish kids, and that's where Meeri Haataja would hang out, too. In fact, so much so that she even found her husband through IRC. Back then the musically talented young woman didn't know it, but technology would become her profession.

In business school, Meeri majored in Quantitative Methods which is mostly mathematics and statistics. She became familiar with data analytics and was hired to do data crunching by the national mobile operator. After several years of analyzing data and finding efficiencies for business processes, she moved on to lead these projects at Accenture where she stayed until 2016.

114 Gordon Kelly, Finland and Nokia: an affair to remember, Wired, 04.10.2013

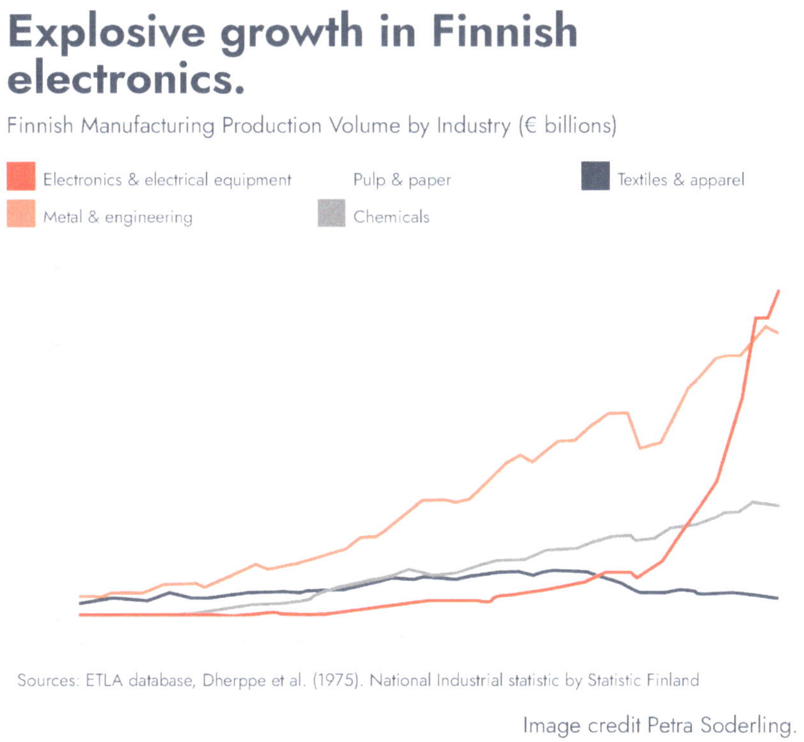

Finnish electronics industry experienced explosive growth in the 1990s.

In 2017, the Ministry of Economic Affairs and Employment commissioned an Artificial Intelligence (AI) strategy for Finland, setting the target for the country to be the leader in the application of AI. The vision for 2025 is to prepare Finland for the age of artificial intelligence and make it into a nation that is competitive and able to attract talent, has the most relevantly educated population, and be a nation whose citizens are well-informed and independent.[115]

The working group that consisted of government, research, and business people, broke this task into four questions:

1. *How can the public and private sector cooperate so that companies have enough support in creating AI-based innovations?*

115 Suomen tekoälyaika, Suomi tekoälyn soveltamisen kärkimaaksi: Tavoite ja toimenpidesuositukset, Publications of the Ministry of Labor and the Economy, 23/10/2017

2. *How could the data owned by public authorities be reused in data-based businesses?*
3. *How does AI impact the future of work and the individual?*
4. *What is required from the public sector on our way to the age of AI?*

This 2017 kickoff report looked at the state of AI globally, Finland's strengths and weaknesses, and also the realistic opportunities for Finland to thrive—both by contributing and receiving—internationally. It also established early on that AI has a profound impact on society in that it touches everyone, and this impact can only be addressed by public-private partnership. The fact that this work is driven by the Ministry of Economic Development, and not, say, the Ministry of Science is a kind of giveaway if you like.

A year later, in early 2018, Meeri was invited to be part of the Finnish government's AI strategy work. Her first role in this public-private partnership was as a member in the accelerator working group, but the group soon recognized a stronger emphasis on the ethics of AI is needed. Meeri Haataja was appointed to lead that effort. Meeri considered it an honor and contributed significant amount of work to the cause.

The Alliance arrived at eight recommendations for going forward: Increase the competitiveness of companies, utilize data in all areas, speed up the adoption of AI, ensure talent, make bold choices and investments, build the world's best public services, create new types of cooperation models, and elevate Finland into a trailblazer.

In January 2019, Meeri's AI ethics group hosted an open workshop to gather comments on *Draft Ethics Guidelines for Trustworthy AI*. The group challenged Finnish companies and government agencies to an Ethical AI Challenge which Finnish businesses were happy to join. In fact, many companies even defined their own ethical principles and guidelines during the process. *Trustworthy AI* was bubbling in Finland.

In 2019, this working group published its final report.[116] In just two years, these eight recommendations had grown into eleven key actions that the

116 Steering group and secretariat of the Artificial Intelligence Programme, Leading the way into the age of artificial intelligence, Publications of the Ministry of Economic Affairs and Employment, June 12, 2019

country now needed to take (was already taking) in order to achieve a vision they set for themselves for 2025. For each key action, they listed achievements to date, lessons learned, and recommendations to move forward.

As each of the eleven key actions contain multiple current and future sub-activities, its web of measures is too vast to depict here. Instead, I will list a handful of initiatives that are directly related to innovation, economies, and industries.

First, the Ministry of Finance initiated a program called "AuroraAI" that helps citizens and companies by suggesting public services to them based on their needs. This initiative supports the goal of creating the best possible public services.[117] In the first stage, the program identified key life events and business activities. Around each event—events as diverse as a private citizen's health issue on the one hand or a need for a business to hire more employees on the other hand—they are creating an ecosystem that consists of key service providers for that particular event. Next, they defined who needed to be part of this planning exercise and created a playbook and guidelines for how to proceed. Finally, citizens and businesses will be able to use the service as a chatbox, a recommendation engine, or other system that is integrated throughout the public administration.

A midterm report of AuroraAI early 2021[118] describes how the program is well underway technically. A fully government-owned company is building the backend AI and network efficiently and with relatively low costs. However, the report stressed that unless more agencies, counties, and communities join the effort and link their data sets, the offer itself and user experience will not be optimal. The program runs until end of 2023, keeping fingers crossed for the Finnish people.

The second pillar, in addition to public AI services, is research. As we discussed in the background piece, Finland started boosting its scientific capabilities in the 1950s and onwards. Finnish AI pioneer Teuvo Kohonen introduced Self-Organizing Maps, sometimes called the Kohonen maps, in 1982.[119] This led the way for larger research in machine learning, neural

117 Implementation of the national AuroraAI programme, Ministry of Finance Finland
118 AuroraAI-ohjelma, ARVIOINTIRAPORTTI, 10.1.2022
119 Self-organizing map, Wikipedia

networks and pattern recognition in Finland. Today, high-level research and education in machine learning and AI is conducted in several universities across the country, as well as at the VTT Technical Research Centre of Finland.

Basic scientific research in Finland is funded by the Academy of Finland, an agency operating within the Ministry of Education and Culture. The Academy of Finland has six flagship programs, one being the Finnish Center for Artificial Intelligence (FCAI). In case you are interested, the other five are working on 6G telecommunications network, materials and bioscience, digital precision cancer medicine, social studies, and photonics.

FCAI comprises 60 professors and 300 researchers and has a €250 million budget for the flagship term of 2019–2026. It brings together top talent in academia, industry and the public sector to solve real-life problems using both established and emerging AI. FCAI's research mission is to create a new type of AI that is more data efficient, trustworthy and understandable. FCAI research programs include[120] Agile probabilistic AI, Simulator-based inference, Next-generation data-efficient deep learning, Privacy-preserving and secure AI, Interactive AI, Autonomous AI, and AI in society.

In addition to FCAI, the Ministry of Education and Culture is also responsible for another key program, the "Elements of AI." In addition to education provided through schools and universities, the Elements of AI massive open online course is tailored to anyone interested in learning how AI impacts their life and work. When this course launched, the target was to educate 1% of the Finnish population or 55,000 people. Today, over 750,000 people (40% women) from 170 countries have passed the course.

In addition, major universities offer more than 250 individual AI courses, 40 master-level programs, 19 bachelor-level programs and 3 doctoral programs. The universities of applied sciences provide an additional 26 study programs on the subject. Overall, about 6,300 students take at least one course in AI at Finnish universities every year. Several hundred students graduate from AI programs annually.

The third leg, in addition to public services and scientific research, is commercial activity. Large Finnish companies were ahead of the curve

120 Artificial Intelligence from Finland, This is FINLAND

and established their own AI research and development centers early on. Where government action makes an impact is with the hundreds of innovative startups with their roots in universities and research. These innovators are successfully applying AI in various business verticals, most commonly in healthcare, manufacturing, business tools and services.

Business Finland, the government organization for innovation funding, investment, trade and travel promotion, funds growth companies in the field of AI. Between 2018 and 2021, 347 R&D projects in over 300 small- and medium-sized (SME) AI companies received €235 million in total funding.[121] Half of this money came from Business Finland, i.e. the government. The other half was covered by participating companies and their research partners. This Business Finland program was meant to support an enterprise-driven partnership model among companies, research organizations and public actors; these programs aim to find solutions to global market disruptions and create new growth sectors in Finland. The results are promising. In 2017, before the AI program began, Finland recognized only 20 startups that focus on AI, as compared to the over 300 SME companies at the end of the program. Exports by the companies participating in the program increased 2.5-fold, valued at €280 million. In other words, the just over €100 million public funding more than paid for itself just by the increase in export value directly resulting from the program. Add the number of new startup companies, the number of existing SMEs starting to create AI solutions, the number of new jobs, the amount of new revenue and taxes, topped with the positive ripple effect in the economy and society—and you have success.

Meeri Haataja is not only contributing to the government's AI Strategy and public conversation about the ethics of AI. She is also the founder and CEO of Saidot, a startup that automates enterprise AI Governance for trust and compliance. The product is used by businesses, governments and public offices. As examples, the cities of Amsterdam and Helsinki are transparently letting their residents know how the city's AI-based tools were created, who created them and how they work.

---

121 AI Business Program Boosted The Development, Growth, and Internationalization of Finnish AI Companies, Business Finland, 18.05.2022

Saidot participated in Business Finland's AI program from the beginning. "Business Finland has been instrumental to the success of my company. Their R&D loan of €300,000 helped us create the first version of our product, and other financial support later to expand and improve it. Business Finland has also visibly promoted us and connected us to international networks of collaborators and clients, this has helped our Go To Market strategy generating interest and inbound leads" she told me. 80% of Saidot's business comes from outside of Finland.

In addition to publicly funded support for AI companies, Technology Industries of Finland, a private industry association, has launched its own AI accelerator. First AI Accelerator (FAIA) helps existing companies to speed up their deployment of AI. In conjunction with the government's strategy, between 2017 and 2020 FAIA was able to more than triple the number of Finnish legacy companies actively using AI in commercial activity from 0.91% to 3.15%. Sixty percent of these companies said they are also developing their own AI solutions, not just implementing third party technology. These companies ranged from energy, healthcare, retail, aviation, maritime, pulp and paper, media, education, real estate, and so on. You might think that 3.15% of companies using AI is a low number, but what's important for our context is to connect the dots from the government strategy to this three-fold growth.

So what does this Finnish story teach us?

According to Meeri, the public-private partnership in the AI working group, and more generally in Finland, is working well. As the leader of the AI ethics group representing private business, she had a counterpart from the Ministry of Economic Affairs and Employment. Meeri felt she and the group has had full freedom to steer the work towards a business-friendly direction. Her partner from the public sector gave her full support. This is also how the general dynamic in Finnish public-private partnerships tends to work. Businesses are invited to share their aspirations and concerns, and the government arm is there to create incentives and to remove obstacles.

During the work for Finland's AI Strategy, Meeri Haataja took part in two overseas trips to the White House to discuss and align policies on digital transformation. Creating those networks with intelligent colleagues in Washington DC was a highlight for her. As Finnish-American technology professional, I'm obviously very happy she did.

As for the future, Meeri Haataja intends to remain an entrepreneur in the field of AI. 2022 and 2023 are the years when new generative AI tools started to flood the market. Many welcomed them, but equal numbers are puzzled by their lack of trustworthiness and transparency. How do they work? Where did they get the data? The potential impact of these tools on our societies, and on how we work and live is mindboggling.

One of the more surprising challenges that Meeri lists is that AI is not used enough. If AI tools were produced and scaled up on a commercially sound level, they would be of a higher quality. Higher-quality products mean more transparency, more regulation, more compliance to jointly-agreed norms and standards. She remains optimistic that these issues will be resolved as AI becomes a norm.

Towards the end of our interview, she says something that stays with me for a long time. We're seeing constant confrontation between innovation and ethics. There are voices out there who try to make us think that regulation is hindering the development of new ideas, new products, new invention. But why would we want to create inventions that would make our children's lives worse rather than better? Meeri Haataja, the European Union, and many others are more inspired by the view that innovations flourish when they are supported by guidance and direction that will benefit all, not few.

KEY TAKEAWAYS:

1. Even an arctic land with a small population can make its economic mark domestically and internationally with smart strategies.
2. Choosing a narrow spearhead scope helps become the best in the world.
3. Adequately resourced public funding will bring results.

ITALY AND THE SPACE INDUSTRY

Leonardo da Vinci, Galileo Galilei, and Christopher Columbus. These people were connected by curiosity, relentlessness, and the tendency to push boundaries. There's another group of less known Italian scientists who were also pushing the boundaries throughout the 20th century. Giulio Costanzi, an officer of the Specialized Engineers Corps wrote a paper in 1914 on the many obstacles for space travel, including navigation and power. His suggestion was nuclear propulsion. "After the conquest of the air through aircraft, it is high time to abandon Earth and found new colonies in space," he wrote. In the 1920s, Luigi Gussalli invented the double-reaction jet engine and ideated moon travel with solar-radiation-powered ships.

After World War II, Gaetano Arturo Crocco invented a liquid-fueled combustion chamber, and Aurelio Robotti built Italy's first liquid-fueled rocket, the AR-3. The AR-3 was launched on May 9, 1952. Mr. Robotti described the event as follows: It "flew straight up for hundreds of yards before disappearing behind a hill." If you do an online image search for "AR-3 and Aurelio Robotti," you will find enchanting pictures of said hill with cows on pastoral land in the Italian Alps. It's fair to say Italy is delivering on the historical goods.

There's an additional name that I'd like to bring to the mix from the 21st Century and that is Luca Rossettini, who kindly agreed to sit down with me and chat about the Italian space economy.

When 5-year-old Luca was listening to his uncle explain about the stars, planets and the universe on a remote, dark Italian mountain top, he didn't understand much of it. But it wasn't the intellectual understanding that was important to Luca, it was the captivating, ravishing feeling that there was something out there that needed his attention. He just had to get himself up there. Luca was going to be an astronaut.

At school, he did well in subjects like physics, mathematics, and other sciences. He also tried out many different sports, being good at most of them but not exceling in any one of them. Young Luca Rossettini was delivering on the plan emerged on that mountain years earlier. There's a once in a lifetime chance to be an astronaut and he was preparing himself on every level. His career took him through parachuting in the Italian army, getting a master's in aerospace engineering, working in a US research lab on nanotechnologies applied to space propellant, and finally completing Ph.D. studies in Advanced Space Propulsion.

After a Fulbright scholarship and NASA training in the US, Luca returned to Europe to found his current company, D-Orbit. D-Orbit is a market leader in the space logistics and transportation services space. You can think of it like FedEx or UPS where instead of trucks, ships, and airplanes, it operates satellite carriers. These are vehicles that can transport satellites in orbit and release them individually into distinct orbital slots.

The modern era of space research began in the 1960s and was inspired by breakthroughs in the Soviet Union and the United States. Italy got involved with the San Marco program which launched five satellites between 1964 and 1988. It might not be well-known that Italy was the third country in the world to go into space. San Marco was a joint operation between the Italian Space Research Commission and NASA. Italy built the satellites and the launch pad, and NASA launched them with American Scout rockets from the Wallops Flight Facility in Virginia. The satellites were doing research measuring general conditions in orbit, and later exploring the relationship between solar activity and meteorological phenomena.

Experiences from building these payloads, launches, and collecting and analyzing data were used not just by the Italians, but for building space capabilities for all Europeans in the 1970s and 80s. The European Space Agency (ESA) has two predecessors that both have Italian connections. The European Space Research Organisation (ESRO) was founded in 1964 by ten nations: Belgium, Denmark, France, Germany, Italy, Netherlands, Spain, Sweden, Switzerland and the United Kingdom. In today's startup vocabulary, you could call it a spinoff from CERN, the European Organization for Nuclear Research. ESRO's roots can be traced back to Italian nuclear physicist Edoardo Amaldi who, together with his French friend Pierre Victor Auger, was a founding father of CERN and instrumental in launching the European Preparatory Commission for Space Research in 1961. This eventually led to the establishment of ESRO.

While ESRO was purely a scientific organization, its sister European Launcher Development Organisation (ELDO) established a vehicle for European satellite launches. ELDO was set up by Belgium, Britain, France, Germany, Italy, and the Netherlands and it funded ten launches between 1964 and 1970. These space launches took off from the Royal Australian Air Force's Woomera Range Complex in Southern Australia, and the ESA space center in Kourou in French Guiana. You can tell how international the space sector was from very early on.

If you compare the timelines, you notice how Italy was simultaneously building its own domestic space industry with the help of NASA and contributing to the European space ecosystem as part of ESA. In fact, this playbook adopted in the 1950s and 1960s—riding two horses at once if you will—is in full use by many European countries today. We'll come to this in a moment. I could imagine the Italians were thinking with the space sector something similar to what Canada is seeing with quantum, "It is premature to pick winners at this stage in quantum's development, but large investments are required." Italy didn't pick winners in space research but worked with the best players across Europe as well as the United States.

Italians can be proud of their achievements. We couldn't imagine space exploration, earth observation, or telecommunication without

ESA and the Italians. Artemis, the International Space Station, the Webb telescope, missions to Mars, these are all joint efforts between ESA, NASA, and other international space organizations. Italy has designed and built 50% of the modules in the International Space Station ISS. Seven Italian astronauts have flown in space. And numerous Italian components, engines, robots, and software programs power dozens of missions. Not bad for a relatively small Mediterranean country.

I asked Luca Rossettini if his decision to found D-Orbit specifically in Italy was due to patriotic reasons, or perhaps the Italian government's support in this sector, and his response was nothing I would have expected—although I should have. Let me back up a little bit. At this point of our conversation, I had realized I'm not talking to your average startup founder and CEO. Descriptions like "Renaissance man" and "Italy's Richard Branson" had crossed my mind. Not least because they both like extreme sports like parachuting, scuba diving and kite surfing. Rossettini somehow manages to talk about boring stuff like international geopolitics and EU funding while keeping that boyish sparkle in his eyes—the same one that lit up some forty years earlier on that mountain with his uncle.

Here is how Luca responded: "I'm not a fan of boundaries. When you look at the earth from space, there are no boundaries. I used to read Konstantin Tsiolkovsky, a 19th century physicist who pioneered in astronomic theory. He was also a philosopher, laying out his vision for how the human race would one day build satellites, put space stations in orbit, and move into settlements all around the solar system. But before doing all that we need to come together as one nation. This is what I believe. Today we are divided into so many nations, defending ourselves against each other. It's all so unnecessary; we're all humans."

Despite Tsiolkovsky, however, Luca is proud to be Italian. "Ours was the third country to launch a satellite after the Soviet Union and the United States of America. Today it's one of the few countries that has a space industry covering the entire value chain." There's a lot of creativity, art and culture in Italy, but also a lot of scientific expertise and technical capability. When combined, these qualities work together like magic. Rossettini does point out that Italians and Europeans in general still have a different mindset from Americans. Europe has been slower to push commercial solutions, and is bigger on regulation, but these things are changing. Europe is overcoming a psychological barrier to change.

The Italian Space Agency (ASI, Agenzia Spaziale Italiana) has been funding, regulating, and coordinating Italian space exploration since 1988. Headquartered in Rome, ASI employs about 300 people and receives its funding from the Italian government. In 2020, this was €1.1 billion,[122] one of only nine countries in the world with a budget this size.[123] Part of this budget (€682 million to be exact) was directed towards ESA, and part allocated to national and bilateral activities.[124] Italy is comfortably the third largest contributor to ESA after France and Germany. Together with the UK, these four countries fund the large majority of ESA's operations. ASI has been receiving a steady financial support from the Italian government (which continued during the Covid pandemic).

In addition to coordinating public activities, the Italian Space Agency ASI is also responsible for boosting the Italian private space sector by creating strategies and suggesting public policy. ASI hosts the website *italianspaceindustry.it* where it reports being one of the few in the world to cover the whole space supply chain, including "large System Integrators and large companies specializing in complex system integration, about 200 small and medium-sized companies engaged in upstream and downstream activities, start-ups able to innovate and stay ahead of new technology trends." Playing with their nifty online catalog is fun. You can search for companies based on size, focus, or location. In 2020, these companies generated €2 billion in revenue[125] with over 7,000 people employed.

Europe and Italy are transforming their space sectors from government-driven to commercially-driven models. For 70 years, space was the playground for governments only, but the past seven years have seen a

122 Annual Report 2020, Agenzia Spaziale Italiana, The Department for Image and Communication Enhancement, Directorate for Communication.

123 Federico Borsari, Aligning US-Italy Relations in Earth Orbit, The Center for European Policy Analysis (CEPA), June 15, 2022

124 Italy, The Space Economy in Figures, The Organisation for Economic Cooperation and Development (OECD)

125 PARTE PRIMO SPACE, IL PRIMO FONDO ITALIANO CHE INVESTE NELL'ECONOMIA SPAZIALE, European Commision, July 27, 2020

dramatic change. Now the large majority of satellites are commercial. This change has also happened in the investor world. Earlier, Venture Capital companies were not eager to fund space startups, since their customers were mainly governments who would offer flat and fixed margins around 8% or as low as 5%. Being dependent on reliable and predictable, but penny-pinching customers doesn't fit the risk investor's model very well!

Recently ESA established a Directorate for Commercialization to boost productization in the European space sector focusing on three key ingredients: talent, access to capital, and fast innovation. According to Luca Rossettini, the Directorate is run by a team of accomplished and smart people who are already delivering results. As an example, ESA now has an advisory board that is populated with representatives from private industry, including the new agile players of the new space economy. This helps the EU bureaucrats get more information on what's going on in the private market, and how best to support that trajectory. As Luca points out, it's not that government employees would make poor decisions out of malice, it has just been that there used to be a lack of communication and real information.

There is a paradigm shift in the private markets as well. The traditional space company was a business whose main customer was their national government. These large multi-billion dollar companies were often serving both the defense sector as well as the space program, and the various administration bodies often only talked to these trusted partners. Neither of these groups would have their fingers on the pulse on what's going on in the startup world or adjacent new space innovations. The change that's occurring rapidly is that not only are there new companies entering the market, but the established companies are changing their ways, too. New business models and new customers are making space exciting for everyone.

Luca Rossettini points out though, that when a market is changing as rapidly as the space market is, the market is also at its weakest. This is why, having those government contracts and that tax payer funded stability is a positive thing. As an example, he told me that when he returned to Italy from the United States to found D-Orbit, he swore never to work with the government again. And he admits that was a big mistake. When a young startup is trying to sell their services, there will be questions like "Okay, then, have you ever flown in orbit?" or "Have you ever done anything according to stringent quality standards?" Because these customers themselves invested millions or thousands of millions into the missions to launch to orbit, they want to make sure their partners are up for the job. Having those testimonials and credentials from previous government launches is important, even when those projects might be slow, and your startup doesn't make a lot of money. Still in this moment of the space industry you need to have both. "A commercial vision that could be enabled by cooperating with the government", as Luca puts it.

In addition to ESA starting their commercialization efforts, things have been progressing with the Italian government as well. When Luca Rossettini started with D-Orbit back in 2011, the Italian Space Agency was reporting to the Ministry of Education and Science. A few years after, a new law was passed which put the agency under all of the ministries, recognizing that space is not just about research, it's as much about business, jobs, the environment, foreign relations, defense, and finance. The debate continues in Italy about the role of space research, space commerce, and other of its aspects. Luca is optimistic, "as long as there is debate there is progress, things are moving forward." The Italians are not building space products just for themselves. They recognize that these efforts contribute to the future of the humankind.

A moment ago, I mentioned how some European countries are riding two horses at once, building their domestic capabilities and business ecosystems as well as contributing to and benefiting from pan-European actions. With the space industry, it is a calculated play with the contributions to European Space Agency ESA's research operations using an instrument called geo-return. Here is the description from ESA's website:

> *ESA's mandatory activities (space science programmes and the general budget) are funded by a financial contribution from all the Agency's Member States, calculated in accordance with each country's gross national product. In addition, ESA conducts a number*

of optional programmes. Each Member State decides in which optional programme they wish to participate and the amount they wish to contribute.

ESA's budget for 2021 is €6.49 billion. ESA operates on the basis of geographical return, i.e. it invests in each Member State, through industrial contracts for space programmes, an amount more or less equivalent to each country's contribution.

In other words, the more a country contributes public money, the more their private companies get business deals. The trick to how to maximize your return is in where and how you choose the optional programs, and Italy has been a master at this. Relying on their decades-long experience in building and launching satellites that are equipped with different types of sensors that collect data, ASI has been able to build a cluster around earth observation and situational awareness.

These are both increasingly important fields. Think of satellite-based forest fire detection and mitigation, crops or mineral deposit monitoring, navigating logistics on sea and land, tracking ice or storm movements, and many other fields that benefit from an eye in the sky. Italy has also been a major player in the Artemis program for moon exploration. All of these space programs also need supporting technology such as robotics, automation, and communications technologies.

This was clear from the vision for the future that Luca Rosettini laid out in front of me: D-Orbit has created the space logistics market. Currently they are the only company in that space, but Luca is welcoming others to join—and they are coming. The goal for this market, a realistic and reachable, I was told, is to create an interplanetary space logistics network. This network is capable of transporting goods and people, but also information. When you imagine this interplanetary highway connecting Mars, the asteroid belt, moon and earth, you soon realize this is what it takes to take the space industry onto a next level. Without such infrastructure the space economy would just be stuck on earth.

D-Orbit's vision is for their logistics network to be part of collecting junk satellites in space, recycling, rebuilding them into new orbital stations in space. This orbital manufacturing will fully recycle raw materials so nothing is left to waste. After all, we don't build ocean liners in the desert and then move them into ocean, why would we build orbital stations on earth, Luca points out. Sooner or later this is going to happen: Someone's just calculated that Starlink alone is sending 29 tonnes of satellites into space every single day. When the infrastructure is in place for orbital manufacturing, the launch industry will become a niche market. We're still going to need rockets, but they will be used in a different, more sustainable way.

Luca compares it to the energy market. Utilities used to have the monopoly in generating energy, gradually they have moved towards being a distributor of renewable energy created in nodes around the country, and eventually their role will become just a manager.

By contributing Italian taxpayer money into ESA programs around these topics, Italian space companies have been able to get a lion share from business contracts directly from the same programs. In fact, in November 2022[126] the newly elected Prime Minister Giorgia Meloni encouraged her government to increase participation in the European Space Agency, including in the areas of launchers, earth observation, human and robotic exploration.

KEY TAKEAWAYS:

1. Focusing on key strengths enables you to develop an ecosystem.
2. Betting on multiple horses—running more than one race—can bring windfall results.
3. Partnering with regional and international communities makes everyone stronger.

126 Decode39 – Geopolitical Insights from Italy, November 23, 2022

SOUTH KOREA AND DISPLAYS

In 1979 Yoonchae Cheong was an aspiring young engineering student at the Seoul National University, in the Republic of Korea. Yoonchae majored in electronics engineering, not exactly knowing what it is. *Semiconductor* was just a new word heard in a lecture, and he couldn't imagine a *cell* phone. His family, living slightly above the poverty line, didn't even have a phone line at home. He had no idea of what kind of jobs were ahead, as there were just two companies (LG and Samsung) making electronic products like radios and black-and-white televisions only. The economy had sunk into the worldwide recession and into some domestic instability, and he was somehow at a loss of direction for his life.

The country's president, former General Park Chung-Hee, had been leading the nation with a tight authoritarian grip since 1961, and by 1979 the Korean people had had enough. Student activists and other protesters were demonstrating and organizing for democracy in their country, and the Seoul National University was not immune to these events. Young Yoonchae was not part of the activism, but he became aware of what democracy should look like by participating in big protests several times. The year 1979 was when his eyes opened to both democratic principles and electronics engineering. He felt more interest in telecommunications than in semiconductors, as it was more like "communication between people" for the democratic society.

Park Chung-Hee was assassinated on October 26, 1979 by the country's own intelligence service, but it took another eight years, until 1987, for the Republic of Korea to become a full democracy. By that time Yoonchae had graduated with a Master of Science degree in Electrical, Electronics, and Communications Engineering and was about to move to Texas to work on a Ph.D. in data compression. This was after 4 years of work at ETRI, the Electronics and Telecommunication Research Industry, the Korean government-funded research institution in Daejeon, where he was part of the technical staff and TDX, the digital telecom-switch development team. Thus, his first career was in telecommunications research and development. The government-funded 4-year project was very successful, and the technology was transferred to the participating companies (which included Samsung and LG). In 1987, having a phone at home was no longer a sign of wealth.

As the introduction suggests, The Republic of Korea, more commonly referred to as South Korea, is a country that has gone from being an unstable agricultural economy to a highly industrialized, technologically advanced western economy in a remarkably short time. In the 1960s, its GDP per capita was comparable with levels in the poorer countries of Africa. By 2004, South Korea had joined the trillion-dollar club of world economies. How did they do it?

When South Korea rose to #1 in the Bloomberg Innovation Index in 2016, the United Nations Industrial Development Organization ran an interview with Professor Heui Jae Pahk, President of the Office of Strategic R&D Planning in South Korea's Ministry of Trade, Industry and Energy. Pahk is a former professor of mechanical engineering at Seoul National University, our friend Yoonchae Cheong's alma mater. In 1998 he founded his own company, an R&D-intensive firm producing test equipment for LCDs, OLEDs, photovoltaics, and semiconductors. As the case with South Korea will demonstrate, research and development is a fundamental driver of economic growth and industrial development. In particular, according to Pahk, R&D is a key factor in enhancing the technological competitiveness of a developing country. It helps the country avoid the middle-income trap, from being stuck in the lowest value-added parts of global value chains. The middle-income trap is an unfortunate situation where the country's wages have risen so they no longer compete as a low-cost country, yet they don't have the means to create their own high-value industries. South Korea escaped this trap beautifully.

In addition to economic growth, the country has had remarkable success in leveraging those achievements to significantly reduce poverty. It was only in 2019 that Korea applied to be released from its World Trade Organization (WTO) status as a developing country, and it was in 2021[127] that the United Nations Conference on Trade and Development (UNCTAD) recognized South Korea as a developed nation.

One of the key drivers for South Korea was the decision to invest in a knowledge-driven economy. Due to the fact that the country holds very few natural resources and must import most raw materials and energy, the South Korean people understood the benefits of and bought into this strategy, eventually to everyone's benefit.

A 2017 study "The role of the state as an inter-scalar mediator in globalizing liquid crystal display industry development in South", published by the Korea University in Seoul, looks specifically at how the display industry was instrumental in pulling South Korea up to the list of developed nations and top of the global innovation index. This paper argues that "the role of the national state as an inter-scalar mediator was crucial in coordinating localized growth factors with globalizing external factors to create and develop the LCD industry".

Let's see what this complicated statement means in practice.

The Korean government used the legacy of the developmental state as a springboard out of poverty. They could have, based on economic success, applied to being categorized as a developed country much earlier than 2019. They didn't do it because being classed as a developing country brought certain benefits: International economic development organizations such as USAID and the World Bank funded and supported Korea's economic growth. These international funds and programs, the "globalizing external factors" had essential impact as they were implemented top-down by the South Korean government.

---

127 Gil Yun-hyung, UNCTAD classifies S. Korea as developed economy, The Hankyoreh, July 5, 2021

Top-down implementation of any program in a democracy, even a young one, requires diplomacy. By being the coordinator and mediator between people with contrasting opinions, the central government brought everyone around the same table so agreements could be made, and implementation could go forward.

From the 1950s onwards, the Korean government had used five-year plans as a mechanism to catch up with western economies. This is a common tool in socialist countries, still successfully used in the People's Republic of China. A good example of the power of these plans is the role of banks. Under their five-year plans, South Korea nationalized banks, and this way was able to channel controlled capital to large, family-owned companies, "chaebols", such as Samsung, LG or Hyundai. Private companies were highly incentivized to implement plans that the central government considered important, including R&D. After the banks were again privatized and five-year planning was abandoned in 1992, the government continued to encourage private companies to learn, adapt, and implement technologies from leading nations including Japan and the USA.

After graduation, we've said that Yoonchae Cheong went to work for the national lab ETRI, the Electronics and Telecommunications Research Institute. As most of the country did not even have landline telephones, ETRI had been tasked to develop a digital switching system to enable building phone lines. The product they were building, TDX, time division exchange, was a success and it helped the country build critical telephone infrastructure. Two other successful publicly-funded projects at that time were the CDMA cellular network standard, and semiconductor hardware. Core scientific development was done in the government lab, but its commercialization was handed over to private chaebols; the TDX went to Samsung and LG, CDMA also went to those two companies plus newly-established Hyundai Electronics, and semiconductors went LG and Samsung as well. All of these technologies contributed greatly to the digitalization of South Korea which was exactly what the leaders of the country had envisioned.

Between 1988 and 2019, these policies and practices resulted in real gross domestic product (GDP) growth averaging 5.45% annually. This robust performance was fueled by annual export growth averaging 9.27% in the same period. These are good numbers by any industrial standard.

By 2003, the idea of a knowledge-based economy had proven itself, but there was more to be done. Until then, the technology industry had been growing on its own, building on the national digitalized infrastructure and the earlier public investments. The administration wanted more coordination, so they strategized resources around ten selected industries or clusters, one of which was digital TV and displays. These industry clusters were recognized as Korea's Next Generation Growth Engines, a win-win strategy involving large companies (mainly Samsung and LG) and small- and medium-sized enterprises or SMEs. The shift from a catch-up economy to an economy that weaves the creativity and imagination of its people with science, technology, and ICT meant promoting start-ups and strengthening SMEs.

Focusing on the ten industry clusters was not the only initiative. The central government worked with regional leaders to create regional growth engines, awarded tax benefits for angel investors, and even directly funded start-ups. The role of the regions has been the critical third leg in this success story that the Republic of Korea is today. If the central government led and funded core research which digitalized the country and created domestic intellectual property, and the large private companies productized and commercialized those chosen technologies, then it was the regions that made it attractive to build factories, employ workers and get those flat TVs, cell phones, and computers out the door and into international markets.

Speaking of an international reach, although this is a remarkable story of one Asian country there are always good friends in these stories that help the protagonist succeed. In our case, it was *co-opetition*—the co-operation and competition between LG and Samsung with their rivals, companies like Sony, Nokia, or Philips. Building a whole new industry takes more than a handful of OEMs. You need a host of reliable partners including suppliers, subcontractors, distributors, marketers, and all kinds of professional groups that will lift the tide and enable this new business to succeed with continuity. As other OEMs were picking up on the same technology, they too invested in the cluster areas. Even better, these international competitors were also supported by their own governments making the cake even bigger for everyone—and serving more people in the process.

For the Koreans to select displays as one of the ten essential industry clusters was not accident. An essential driver in the market was the trend to replace tube-based, thick, fat televisions with flat TVs that would look sleeker and take up less space in homes. People had money and they wanted their flat TVs. Liquid Crystal Display (LCD) was not a Korean invention. The research first started in the US in the 1960s, but Americans didn't have enough interest to make the heavy investments that it would take to get this technique out of the laboratories and onto store shelves.

LCD was the first technology that enabled thin and light displays. It was successful in both creating the market for these types of devices and making lots of money for industry players at the same time. But soon came an even better technology that enabled even better TVs: OLED, organic light-emitting diodes, is based on an organic substance that acts like a diode. The main difference between LED and OLED is that the latter doesn't need separate background light; LCD requires a fluorescent tube, or nowadays an LED light.

Takatoshi Tsujimura, a Japanese-born scientist who has been credited with commercializing OLED technology, remembers the early days like this:

> *"The basic structure of organic light-emitting diodes (OLEDs) was reported by Tang and Van Slyke at Eastman Kodak in 1987. Before any in-depth discussion of OLED display structure, let us consider the initial origins of OLED technology, which are based on early observations of electroluminescence (EL). In the early 1950s, a group of investigators at Nancy University in France applied high-voltage alternating-polarity fields in air to thin films of cellulose or cellophane containing deposited or dissolved acridine orange and quinacrine, and observed light emission. One mechanism identified in these reaction processes involved excitation*

of electrons. Then in 1960, a team of investigators at New York University (NYU) made ohmic (a nonrectifying charge injection, which shows linear current– voltage relationship) dark-injecting electrode contacts to organic crystals and described the necessary work functions (energy requirements) for hole and electron-injecting electrode contacts. These contacts are the source of charge injection in all present-day OLED devices."

The involvement of universities in the background research that led to breakthrough inventions in displays largely happened outside of Korea; South Korean companies, with the help of their government, have been phenomenally successful in implementing these technologies in consumer electronics products. Who doesn't have a Samsung or LG product in their home?

After the boom days of explosive success of displays, semiconductors, and telecom products, the public officials took a step back and allowed the market to play its course. It hasn't been until recent years that the South Korean government started to consciously increase the nation's R&D spending. As South Korea is no longer a catch-up economy only concentrating in rapid adoption of technologies from other countries, it has recognized the need to stay at the top by relying on homegrown intellectual property. South Korea's R&D expenditures reached $75.4 billion in 2020, accounting for 4.81 percent of GDP. That made South Korea the second biggest R&D investor in the Organization for Economic Cooperation and Development (OECD) countries trailing only Israel, whose R&D investments came to 5.44 percent.

According to the OECD, South Korean's biggest strengths are the willingness to adopt new technologies and the high number of educated young people in the fields of engineering and science. More young people study these subjects in South Korea than in any other OECD country. Most universities provide a comprehensive subject base covering science, social sciences, the humanities as well as engineering and business. Although having stellar education outcomes linked to a phenomenal economic success, it might surprise you that universities in South Korea are largely privately funded. The Korean government invests heavily in education, but most of this funding has gone to providing basic education for everyone, creating the base for knowledge economy. In recent history, this strategy has been lifting its people out of poverty.

Universities have not been, and likely will not become, major players in the national innovation field for the foreseeable future, as a great deal of current industrial R&D is going into developing new products and markets. The role of universities is to do basic research, publish scientific articles and studies, and to train future researchers. Today, their national labs share these goals. Innovation is left to private industry.

By looking at the South Korean display industry, its history and success, we are revealing a government-encouraged innovation model that seems to be working for everyone. The state has played a key role in building the enabling digital infrastructure, educating its people, creating critical intellectual property, incentivizing private companies to productize, and regional governments to manufacture. The post-1987 government also realized when they were needed and when they would be in the way and knew when to graciously step back. What's not to like?

What became of our young student friend Yoonchae Cheong? After working at Hyundai Electronics on the CDMA product, he built an international career in telecommunications, working at Nortel Networks and Samsung for a number of years. After building products in private industry, Yoonchae transitioned to the National Research Foundation (NRF) of Korea. His responsibility was to execute the R&D budget of government by making Request for Proposals (RFPs) and by selecting grantees. The technologies he covered include quantum computing, supercomputers, and convergence technologies among all tech sectors. That experience opened up a new road to Washington D.C. to facilitate the international cooperation between USA and Korea in quantum computing technology.

Would it surprise you if I said that one of the newest crazes in display technologies is something called quantum dots? People like Yoonchae will ensure Korea remains successful for decades to come.

KEY TAKEAWAYS:

1. Top-down governance works when a country needs a major overhaul.
2. Government can help by creating infrastructure, education, and investing in domestic intellectual property.
3. After incentivizing private companies, the market can run on its own.

INTRODUCTION TO SECTION 3

Thank you for following the journey thus far. We have now established that hope and optimism work best when you want humans to achieve new goals, that there is a strong correlation between regulated countries and successful countries, that public investments yield a better return than private investments, and that one person can make a change.

The purpose of this third section is to convince you that you are this one person, and to show you some of the tools that can get you started.

My ambitious goal with this book is for it to be the book that has never been written before. Not just as a single book, but as a category of books speaking to an audience of Economic Development professionals. This is a group of people that tirelessly and selflessly work for their communities, often creating a large economic impact that leads to jobs, tax revenue, and a better environment for everyone. Equally often, I'm afraid, it is a group of individuals that is unrecognized, undervalued, under-appreciated.

When I started doing Foreign Direct Investment work, no one told me how it should be done. I did have a generic training. I had to figure out for myself things about the specifics of my two regions, the source and destination countries, and the focus industries. I had to determine what the offering from the receiving country is, what the need in the source country is. I had to understand what companies in these regions do well and where their needs are. Later, as I did economic development work with certain US states, I ran into the same issues. There is very little material on "how to do" economic development—and do it well. I would have appreciated a handbook of sorts to guide me through the strategies I ultimately found valuable and realistic expectations of how things would unfold.

I hope this next section, this call to action, can be considered a sort of handbook for economic development professionals who are interested in their trade and want to improve what they are doing. I hope it helps those who most importantly want to know how to build long lasting success stories that burst with prosperity, intellect, balanced lives and happiness in our communities.

SECTION 3

CALL TO ACTION

HOW TO MAKE A CHANGE

In this third section, we will discuss how to use the tools at our disposal, inspired by the example countries to make a change in our own environment.

My suggestion is to start by thinking about what kinds of problems there are in the world that need immediate solving. These problems are large-scale issues that lead communities, organizations, researchers, business people and the public policy makers to one or more working solutions.

Investing in solving problems that are very big will always yield a return and, if not the full solution, at least you will be part of a bigger trend toward a total solution alongside the others working with you.

Then, we will look at your country, state or city. What kind of resources do you already have? Is there anything there that could be further developed into an industry that is future-proof and will result in economic growth?

Finally, we will bring it all together for you so that you walk away with a working hypothesis.

Sometimes when you look at the world or you think about the economy, it seems overwhelmingly complex. Complexity, or the appearance of complexity is a given that we just need to accept. It might feel overwhelming to think that, if I now do this act, let's say suggest a new law that adds entrepreneurship as a new school subject in middle-school, what real impact will it have? The chain of events that follow seems way to random and impossible to control. How am I going to prove to my voters that it will work, or in the future that it has worked? I cannot answer that questions, but I might be able to offer some in the form of the theory of universality.

The theory of universality[128] presents a new force in nature, the universal force. In this theory, everything in the universe, be it space, time, matter, energy, are bound together in an inter-relationship, held together by this universal force. On one hand, the universal force is fighting against randomness and chaos, bringing the element of predictability into systems. On the other hand, universality is also fighting complete order, bringing the element of unpredictability into systems. Therefore, according to universality, our societies are both predictable and unpredictable at the same time.

Universality was first discovered in the 1950s on an energy spectrum of uranium nucleus. This energy spectrum has hundreds of protons and neutrons that move and shake in infinitely many ways yet would always eventually form this same pattern that the scientists called universality. In the image below, you can see the universality pattern in the middle as grey, and how it differs from totally random, the red pattern, and from totally organized, the black spectrum patterns. In 1970s universality pattern was found in prime numbers, and year 2000 a Czech scientist discovered it in bus routes of the city of Cuernavaca in Mexico. The distance and time between each bus, as observed long enough, would not be random and it would not be periodic, but would obey the law of universality.

Today, this same universality pattern has been found in the network nodes of the internet, in a pattern of snow on a frozen lake, in climate models, human bone density, and in chicken's retinas. It is believed we've just scratched the surface what this finding means, but scientists agree the pattern is real.

Coming back to your new legislation proposal to add entrepreneurship as a subject to middle-schoolers, this should bring some comfort.

128 Natalie Wolchover, In Mysterious Pattern, Math and Nature Converge, Wired, February 6, 2013

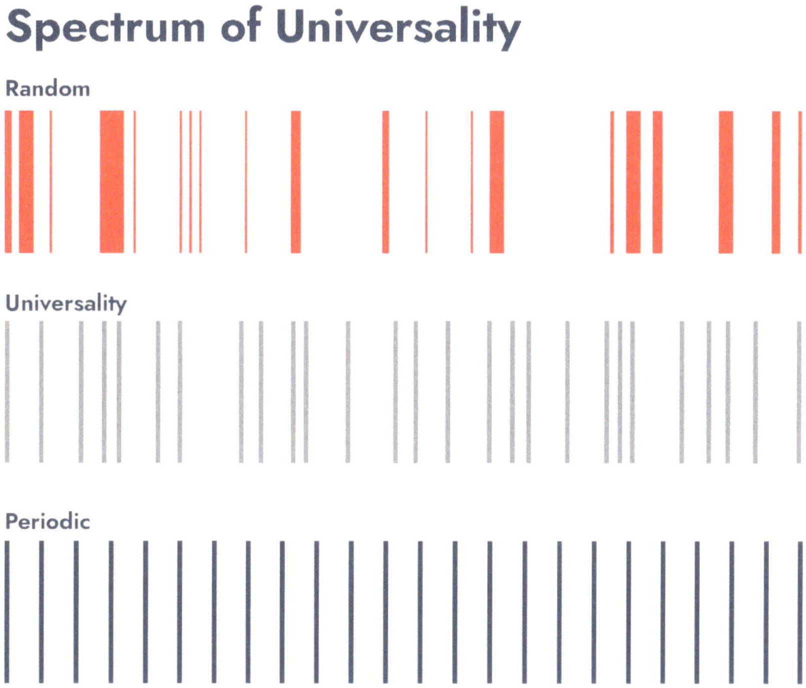

Illustration source Simons Science News. Image credit Petra Soderling.

Nature has a tendency to organize in a pattern of universality, rather than in random or periodically even pattern

Even though the outcome may still unpredictable, we know it is not going to be 100% random, and we know it is not going to go 100% according to the plan. However, it will have a beautiful, rhythmic shape and it's going to make a change.

If this sounds too theoretical, I have a nice story[129] that describes how the duality (the universality) was discovered in ecology.

In the early 1980s, East German student Hendrick Schubert finished his undergraduate degree in ecology in the University of Rostock in East Germany. He moved on to teach in a different university in the West, but some years later was invited back to fill a professorship at his Alma Mater. One day he walked into a lab there and he saw one of his former

129 Life in a Barrel, Radiolab, March 4, 2022

professors, Reinhard Heerkloss. Next to Heerkloss, he also sees, much to his surprise, a bright blue 100-liter barrel. As an undergrad, Schubert had done an experiment with Heerkloss where they had filled barrels with seawater. They were tweaking the nutrient levels to watch how it would affect the microorganisms living in seawater. It was a simple little experiment that had only lasted for two weeks. And now, more than a decade later, Heerkloss still had that barrel sitting there.

Heerkloss tells Schubert how, sometime after the initial experiment, the Berlin Wall fell. What felt like overnight, everything changed. The currency changed, the head of state changed, the university changed its name and its curriculum. The barrels were forgotten. Much later, a colleague of his in his department wanted to do a different experiment and asked Heerkloss to remove the barrels. To comply, he started to empty and move the barrels one by one. Heerkloss continued until he got to the control barrel. That barrel had remained unopened since the beginning of the experiment over a decade earlier. For some reason, right when he was about to tip it over, he stopped. Curious, he thought he would just take a little sample to see if there was anything there. He doubted it, but did it anyway.

As he looked in the microscope he was totally dumbstruck by what he saw. The sample was filled with organisms! The barrel was not just alive. It was thriving. There were green phytoplankton, zooplankton (some of which eat the phytoplankton, some of which eat the other zooplankton). And then there were bacteria that were recycling the whole system.

Unwittingly, he had created a little natural world.

In the 1980s and 90s in the field of ecology, the natural course of an ecosystem, i.e. how ecosystems are born, was an open question. Scientists wanted to understand what would happen if you just gave an ecosystem the basic things it needs like sunlight, space, nutrients, and water, but no outside confounding factors.

You might be remembering our theory of universality here and you would not be wrong. There were two competing lines of thoughts. According to one theory, as all the different creatures get to a certain population level, the level stabilizes and the ecology would form a stable line. The second theory was more like a cycle. Here, there would be one species that would dominate for a while, but eventually it would crash and another species would take over. The prevailing view among scientists was that when left alone, nature tends to get balanced, i.e. form a stable line.

Heerkloss thought he had the perfect opportunity to test the theories. Here, he had an ecosystem that was totally untouched by humans. And the species in that ecosystem are of a type that are born, reproduce and die at a super-fast pace. So in just a few months' time, he could see hundreds of generations worth of transformation. Heerkloss started tracking how the various species were doing. For the next six years, he watched over this tiny universe where generations were passing in the blink of an eye for him. He was waiting for harmony—that stable line—to arrive but it never did. There was no line, and there was no cycle. In a nutshell, there was only what seemed like chaos to him.

With Hendrick Schubert back in Rostock, together with Heerkloss, theoretical ecologist Elisa Bennica, and other scientists, they jointly published their theory in Nature Magazine. Bennica's contribution to the paper was to address the seemingly chaotic outcome. Bennica clarifies that chaos is not randomness but a system, which is high in predictability in the short-term, but cannot be predicted in the long-term. Weather is an example of this. Bennica's explanation was met with immediate pushback from other ecologists. The scientific community feared that if ecology was just chaos, why continue doing research? Why try to restore nature in any way?

Hendrick Schubert, skeptical scientist that he was, repeated the experiment several times with eight barrels. He found signs of chaos in some of the vessels, and in some of the compartments tested, but not in all eight and not always in the same barrels. Additionally, when there was chaos, it played out in different ways in different barrels. The conclusion in Schubert's second experiment was that he still didn't know for sure if ecology demonstrated order or chaos.

This story nicely supports our earlier theories that nature—our home—can be both deterministic and unpredictable at the same time. In our quest to establish parameters for how to make a change, we have found suitable attributes to our physical surroundings. We are mammals living in nature as part of the ecology and bound by laws of physics. As we set goals and expectations for ourselves these laws apply. Things can go more or less as we planned, or they can go differently (sometimes even better than we hoped).

In addition to being mammals, we are also a herd species. We survive by picking up social cues from our fellow humans and aligning our actions to the coordinated groups. There are many real life and lab experiments where social pressure has made an individual do things they normally would not do. Think of concentration camps, or the famous Milgram experiment where a professor was able to pressure his students at Yale to give what they thought to be lethal electric shocks to their fellow students. Social disapproval provokes our danger circuits,[130] that's why we stress over belonging.[131] Our tendency is rather to be comfortably wrong with the herd than to be correct all on our own.

Living in herds has not only mental health benefits but is a means for survival. The "Many Eyes hypothesis"[132] describes this in a school of fish or flock of birds. With a large group of animals there are progressively more eyes scanning the environment for predators. An individual group member needs to spend less time to be vigilant as the task is spread out. As long as one member of the group receives an alert of danger, everyone in the group receives the alert. But alerts are not the only thing that is shared within the group. Individuals also continuously scan each other in order to determine what level of alertness is required of themselves. This is how even one individual is able to get their message across the entire herd; their message will spread in waves.

As social animals, the Many Eyes hypothesis reminds us to move from an individual mindset to a community mindset if we are to evolve, grow, create and prosper. We are not alone; we do not work for ourselves so why should we work by ourselves? By using the ears and eyes of your community you not only get more input from your surroundings, but

---

130 Noam Shpancer, You Are a Conformist (That Is, You Are Human), Psychology Today, December 5, 2010

131 Steven L. Lima, Back to the basics of anti-predatory vigilance: the group-size effect, Animal Behaviour, January 1995

132 Xinxin Wang, Le Yang, Yumeng Zhao, Cong Yu and Zhongqiu Li, The group size effect and synchronization of vigilance in the Tibetan wild ass, Current Zoology, May 27, 2020

you get the community on your side, with you, scanning for food and danger, i.e. opportunities and risks.

Inventor of the polio vaccine, Jonas Salk[133] suffered from not realizing this before it was too late. He was said to have been an unconventional scientist disliked by his peers even before his breakthrough. Salk was driven by an intense desire to help mankind through lonely days and nights spent in the lab reading, experimenting, and documenting his findings. When he published the polio vaccine, he went directly to the public and basked in the media limelight without giving credit to other researchers whose work preceded his own. This is important, as there is a strict code in the scientific community to credit others where credit is due. Anyone seen to take credit for another scientist's work is disapproved. Scientists also are not big on media attention, so Salk was rubbing many colleagues the wrong way. Being ostracized by his fellow scientists cost Salk and his family socially, and likely cost Salk professionally later on as he was not a welcomed member of the science community. Don't be a Jonas.

In fact, a great read on the power of interaction between an individual looking to make a difference and with others, is Robert Greene's book "The 48 Laws of Power." It is a good read for anyone needing to work with communities to get their ideas accepted and implemented. I will describe three of these laws that are relevant to us as we plan to make a change.

"Be royal in your own fashion (Law 34)." This could sound familiar to those who meditate or use affirmations to achieve life goals. Whenever you are around people (and why not when alone, too?), imagine that you are wearing a crown. This crown radiates regality and confidence and it will help you to ask ambitious things that no ordinary person would dare to ask for. Carry yourself in a way that invites respect and trust. Do this by respecting yourself and by inspiring other people to respect themselves. In a group like this, power will naturally gravitate towards the one wearing the crown, and you will be able to get a buy-in for big decisions much more easily than by trying to force them on a doubtful crowd.

"Preach the need for change but never reform too much at once (Law 45) ." When you're trying to introduce change—even while wearing

133 Jonas Salk, The American Academy of Achievement

the crown—you must respect the traditions and set ways of your fellow people. Intellectually people understand the need for change, but emotionally we are hard wired to resist change. There can be such a thing as too much innovation, especially if it seems it's going to affect people personally. Often people recognize the need for change in the abstract but reject the idea when they have to make personal sacrifices (which you know if you ever tried losing weight, for instance). People can become traumatized by too much change too fast. Respect their experience and offer change as a gentle improvement of the past and the present.

"Plan all the way to the end (Law 29)." In our everyday work we create business plans, strategies, and other time-sensitive actionable objectives. We often think we have the end game well figured out, when in fact many times we are merely hoping for this outcome. This is how we would like for the plan to play out. By acknowledging the complex or chaotic environment, by wearing your crown and influencing your constituency, and by introducing plans that are not too radical and not too ambitious time-wise, you possess the core set of tools to ensure you are not making plans that are based on hope as a strategy. Greene also warns us about "the vultures circling overhead trying to live off the carcass of your creation." Planning all the way to the end is like playing chess and anticipating your opponent's moves several steps ahead. There will always be people who want to see you fail. In the case of success, they will want to take the credit.

Pride, vanity, love, and hate are human emotions that come into play at home and at work in politics, in commerce, in academia, in the military. The only way to make change happen is to embrace our humanness in ourselves and in others.

KEY TAKEAWAYS:

1. You and me, all of us, are living in a chaotic environment.
2. Our togetherness can override the chaos.
3. Be smart in how you define your position in this togetherness.

WHAT THE WORLD NEEDS

The lyrics of Jackie Shannon's evergreen hit from 1965 vocalized that "what the world needs now is love, sweet love. No, not just for some, but for everyone."

The young generation of 1965 was taking action to move the world toward what they felt was a more just, peaceful and more equitable place. They protested senseless wars, racial discrimination, gender-based discrimination, and environmental pollution. They staged countless sit-ins. They marched for peace, love and understanding. And you remember (and if you don't, it is historical fact so look it up!) that they were able to get laws passed that allowed women and people of color to have full human rights. They were also instrumental in the banning of some of the most dangerous pollutants and the pollution clean-ups of the era.

Birth control pills became legal in 1965, multiple Civil Rights Acts were passed between 1957 and 1968, and the first Earth Day was celebrated in 1970.

This was the counterculture of my parents' generation and I'm thankful for their achievements.

Those then-radical, now-normal policies advanced our societies by making our world more comfortable, more livable for the many rather than just the few. Unfortunately, under the law of universality, progress is not linear, so today we find ourselves far from the utopia that the 1960s counterculture generation envisioned. We still face numerous challenges around the globe—and above it!

A challenge we've created above our planet is space junk. Within a few decades, we've seen our planet surrounded by not only thousands of small functioning satellites we launched into orbit but also an increasing amount of space junk that comes from that.

NASA states: "More than 27,000 pieces of orbital debris, or "space junk," are tracked by the Department of Defense's global Space Surveillance Network (SSN) sensors. Much more debris—too small to be tracked, but large enough to threaten human spaceflight and robotic missions—exists in the near-Earth space environment. Since both the debris and spacecraft are traveling at extremely high speeds (approximately 15,700 mph in low Earth orbit), an impact of even a tiny piece of orbital debris with a spacecraft could create big problems."[134]

A Guardian news report in 2022 stated, "RAF Fylingdales in the North York Moors England, or Fylingdales, is part of the US Space Surveillance Network, which monitors about 30,000 items of space junk larger than 10cm. The network issues warnings when objects might collide, hopefully giving operators time to assess the risk and move if need be. But there are more than 100,000,000 pieces that are too small to track and eminently capable of causing damage. Travelling at more than 15,000mph, micrometre-sized particles can chip windows and dent solar cells, while millimetre-sized flakes can destroy satellite cameras or puncture astronauts' space suits. At 1cm and above, a speeding fragment can take out an entire satellite.

And the junk stays in space, if not forever, potentially for centuries:

"According to the UN Office for Outer Space Affairs, had the early Romans launched a satellite into a 750-mile-high orbit, it would only fall back to Earth about now.

The risk that space debris poses to satellites and humans in orbit has not stopped countries adding to the problem when they want

134 Mark Garcia, Space Debris and Human Spacecraft, NASA, May 26, 2021

to flex their muscles. In 2007, China destroyed a defunct weather satellite with a missile, increasing the amount of space debris in low Earth orbit by 30%. The US and India followed suit in 2008 and 2019 respectively, cluttering the environment even more. Russia's own act of orbital vandalism last year [2021] created more than 1,500 fragments that will circle the planet for years."

Guidelines for space debris mitigation and the long-term sustainability of outer space activities, published by the UN Office for Outer Space Affairs, lay out how nations and companies should behave in space. Operators are meant to remove satellites from orbit within 25 years of completing their missions. But not everyone follows the guidelines. While most new rockets are now "de-orbited" once their work is done, the vast majority of satellites in low Earth orbit are simply abandoned to their fate."[135]

While we are cluttering the outer atmosphere, the surface of the earth is also cluttered with chemicals, plastic, and waste: in other words, with pollution. While some pollution affects our human health immediately (see the Erin Brockovich story for an example, well-portrayed in the 2000 film of the same name about ground-water contamination by a huge corporation), we may not know the effects of other kinds of pollution we are producing for some time to come. We must, however, become *predictive* about it. It's possible, as the Brockovich film demonstrated; we just need the will.

Air, land and soil, marine and coastal, freshwater, chemical and waste are all polluted intensively and on a global scale.

"Plastics have been around since 1907 (with Bakelite) but production and the vast new usages of such materials exploded starting in the 1950s.

Around 20% of all plastic waste in the oceans comes from marine sources. The other 80% comes from land. In some regions,

135 *Ian Sample, 'Mind that satellite! The mission to clean up dangerous space junk, the Guardian, 21 Apr 2022*

marine sources dominate: More than 80% in the Great Pacific Garbage Patch (GPGP) come from fishing nets, ropes and lines.

High-income countries generate more plastic waste per person; however, most of the plastic that ends up in the ocean comes from rivers in low-to-middle income countries. This is because they tend to have more mismanaged plastic waste, whereas high-income countries have much more effective waste management. This makes the improvement of waste management systems across the world critical to reducing plastic pollution."[136]

"Estimates by the European Environment Agency suggest that 62 per cent of the volume of chemicals consumed in Europe in 2016 were hazardous to health. The World Health Organization estimates the burden of disease from selected chemicals at 1.6 million lives in 2016. Global chemical sales, not including pharmaceutical drugs, are expected to double between 2017 and 2030. Asia is expected to account for almost 70% of sales by 2030."[137]

"Common waste is dumped in landfills and in oceans and includes construction, industrial, household, plastic, electronic, radioactive, sewage, and run off waste (which itself can include chemicals such as fertilizers or pesticides and run off into groundwater, rivers or oceans). The act of dumping and implementing poor or improper waste management can result in soil pollution, air pollution, marine pollution, and groundwater pollution.

These types of pollution jeopardize human food supplies from land and sea, and are the direct cause of millions of deaths per year. By March 1st, the year 2023 has already registered 1.6 million deaths from air pollution alone."[138]

While space junk and earth/ocean pollution can be detected with the human eye, it's a different story with digital and other types of challenges that creep up on us, sort of like the frog in a pot of cold water that is being boiled.

---

136 Hannah Ritchie and Max Roser, Plastic Pollution, Our World in Data, April 2022
137 Chemical Pollution, United Nations Environmental Programme
138 Global Waste Problem, Global Challenges, The World Counts

We've created invisible digital robots ('bots') that crawl through our social media platforms. Bot developers, owners and operators seek to "mediate" the nature and content of individual users' postings.

The underlying perception is that we post both flammable and supportive political, social, religious, economic, ideological and other content resulting in real world outcomes. We must, however, ask if they are the outcomes we hoped. We need to ask if decades of success in seeking and gaining freedoms and human rights is not being quietly reversed by those who have their own agendas.

Although we are legislating on equality and equal opportunity regardless of age, gender, sexual orientation, religion, or physical or mental impairment, we are starting to wonder if the laws are creating the outcomes we expected.

We also need to ask if we legislating for equality around the world, or just in the developed nations? Inequality has been legislated, too!

> *"... examples of discriminatory laws, policies, and norms exist on all continents, and in every nation. Gendered laws and policies are associated with fewer girls enrolled in primary and secondary education; fewer women in skilled work; fewer women owning land or property; fewer women accessing financial and health services; and higher rates of domestic, family, and sexual violence. Thanks to research from the World Bank and UN Women, we know that sexist laws, policies and norms are pervasive, and their economic, social, physical, and mental impact on girls and women is significant."*[139]
>
> *"Studies in the US are showing that while we can legislate for equality through new laws and educate through new training programs, we cannot legislate a change in personal behaviors. It's like giving a lecture on climate change and expecting all attendees to immediately stop driving their cars to get to work or taking a plane across country. At best, legislation and training create new awareness, but it is as yet uncertain where that awareness might take us.*

---

139 Michelle Milford Morse, 'Gender Discrimination Is Enshrined In Law. That Needs To Change', UN Foundation, March 8, 2022

> *Legislation on equality is not an instant pill, and believing it is might just be like trying to swallow a whole elephant. Training recognizes that behavioral change is a process and requires a commitment. But training won't be taken seriously if the lecturers and facilitators themselves have no training and no science behind their presentations.*
>
> *Like our parents or grandparents staging sit-ins, we must recognize that once-and-done is not an effective approach to change. We have to use all the tools in the toolkit, and commit to implementing what we learn."*[140][141]

No one is shocked that there is still a divide between rich and poor nations (though we should be). Yet, when there are still homeless and hungry children in the world's richest countries—in our own developed, rich nations—we should sit up, take notice, and act. It's long been proven that if we just distributed the food products from agriculture that we are now wasting or otherwise destroying, we could feed everyone on the planet.

When Deloitte attempted to survey homelessness in the US and around the globe, they discovered that there is no single global agency measuring homelessness. They also discovered that each nation (and within nations' regions and states), measures homelessness by different metrics as well as measuring it with different frequencies. Nonetheless they were able to establish that homelessness is still very real in wealthy nations of the planet, and that it is largely an urban phenomenon.

> *"In 2018, 81.7 percent of the United States' homeless were in urban areas—with New York and Los Angeles alone accounting for 26 percent of this number. The situation is similar across the Atlantic in England, where 61,410 households were owed relief under the*

140 Martin Kaste, NYPD Study: Implicit bias training changes minds not necessarily behavior, NPR, September 20, 2020

141 Lybi Ma, 8 reasons why it's so hard to really change your behavior, Psychology Today, July 22, 2017

latest statutory homelessness laws in the fourth quarter of 2018, nearly one-fifth of these were in London."[142]

"Across the world, about 1 billion children are multidimensionally poor, meaning they lack necessities as basic as nutrition or clean water.

Considering that 41 countries that are members of the Organisation for Economic Co-operation and Development (OECD) and/or the European Union (EU) are the "rich" nations, a study determined that "in almost half of rich countries, more than one in five children live in poverty. In many countries, the poorest children are at greater risk of depression, obesity and low academic achievement."[143][144]

We don't uphold the idea of one person-one vote. The concept has a fairly substantial history.

"Since at least 1880, in England, the US, and indeed around the world, the concept of one man-one vote has taken hold, in view to ensuring the equal representation of individual citizens.

Historically, the emphasis might have been on representing areas rather than people, and those who were allowed to represent were often powerful land owners or otherwise wealthy individuals of influence (almost always male), while those "represented" had no vote themselves.

India and South Africa have implemented the one man-one vote system with success. However, where it was in place in nations as different from each other as Russia, Belarus, Zimbabwe, Zambia and Angola, when a strongman (dictator) came to power the equitable one-one voting was ended.

142 Barya, Frey, Suriya, Byrnes: The homelessness paradox, why do advanced economies still have people who live on the streets?, Deloitte, September 15, 2019

143 United Nations Child Poverty Programme

144 United Nations Report: World of Influence, Understanding what shapes child well-being in rich countries, September 2020

Absolute monarchy Brunei has not held a national public election since 1962, so universal suffrage becomes meaningless and there are cases of strongmen in power who do not hold elections.

Today, women now have the right to vote in every nation (except the Vatican), but that doesn't mean they are guaranteed the ability to vote. Suppression of the female vote is easy in patriarchal societies where the male can withhold the female's ID card (needed to vote) or community and family harassment can prevent females from going to a polling place." [145][146]

We can perhaps extrapolate from the evolution of one-man-one-vote that differences in treatment exist elsewhere, and nowhere is this truer than firstly, in incarceration practices and next, how we seem to stand by as dictators of nations commit mass-murder.

We tend to assign prison sentences based on race or financial wealth or type of crime or upon whom a crime was perpetrated. The racial disparity in the US prison population is telling. In the US, it is common to speak of "white collar crime" as versus "violent crime". Wealth also plays a role when we consider the assignment of a Public Defender for free to those accused who cannot afford to pay for a lawyer versus those who have "battalions" of attorneys to defend them.

"The latest available data regarding people sentenced to state prison reveal that Black Americans are imprisoned at a rate that is roughly five times the rate of white Americans."

In the USA, "states have begun to chip away at mass incarceration. Nine states have lowered their prison population by 30% or more in recent years: Alaska, New Jersey, New York, Connecticut, Alabama, Rhode Island, Vermont, Hawaii, and California. This decline has been accomplished through a mix of reforms to policy and practice that reduce prison admissions as well as lengths of stay in prison. Still, America maintains its distinction as the world

145 One man one vote, Wikipedia
146 Countries where women cannot vote 2023, World Population Review

leader in its use of incarceration, including more than 1.2 million people held in state prisons around the country."[147]

Don't believe the US is alone in needing to review its sentencing and imprisonment laws, practices, and policies. The UK-headquartered *Institute for Crime & Justice Policy Research* (https://www.icpr.org.uk/) does international, policy-oriented research on justice.

"The ICPR research team monitors trends in world prison populations and examines the causes and the consequences of rising levels of imprisonment. The Institute's latest comparative research on the sentencing of burglary, drug importation and murder highlights vast disparities among the ten nations in their approaches to sentencing. The ten nations (which actually include regions of two countries) are Kenya, South Africa, Brazil, New York State USA, India, Thailand, England and Wales, Hungary, the Netherlands, and New South Wales Australia.

"For a burglary involving prior offences, you would probably get an immediate custodial sentence of several years, with three or four years likely in most countries. Only in the Netherlands would you have a real prospect of avoiding prison. Note that non-custodial options would have been possible, even likely, in most of the ten countries, without the aggravating factor of your previous convictions. If you plead guilty and agreed to make reparations, you might not be incarcerated at all.

For drug trafficking, you could be looking at life imprisonment and a hefty fine if you were sentenced in Thailand, but be home within months if you happened to be in the Netherlands. You'd be facing a 20-year term in India, 15 years in South Africa, and between 5 and 18 years in most of the other countries.

As for murder—"homicide with intent"—a life sentence would be the probable outcome in most of the ten countries. This would

147 Ashley Nellis, The Color of Justice: Racial and Ethnic Disparity in State Prisons, The Sentencing Project, October 12, 2021

> *currently take effect as a whole life sentence (that is, with no possibility of release) in Kenya (where the death penalty is still on the books for such a crime), potentially also South Africa, and as a minimum term of 20 to 25 years in England and Wales, New York State and New South Wales, with a right to apply for release on parole thereafter."*
>
> *"These data points are only the tip of the iceberg: Disparities in length of sentencing around the world exist; the goal of incarceration is not always clearly stated; reforms to nations' sentencing and incarceration practices are not tried-and-tested on a global platform. The role of wealth and influence might, in some jurisdictions, determine whether the accused "gets off", "gets a reduced sentence" or is incarcerated at the statutory maximum sentence up to and including life sentencing or the death penalty."*[148][149]

Why is it that we stand by—as individuals, regions and nations—when dictators kill people, commit genocide, mass-murder? Do we have so little power as on-looking nations, as individual human beings?

Dictatorships. Repressive political regimes. Genocides. Pogroms. Holocaust. Concentration or death camps. Mass murders and mutilations. Intentional mass starvation. Forced labor or forced marches to exhaustion and death. Any and all of these terms should make our skin crawl.

Yet examples of these has recurred throughout human history, and alas, since the beginning of the 20th century up till today as well. What this means is that in our lifetimes (and that of our parents, grandparents, and great-grandparents—people we *know*), we have been witness to arbitrary mass slaughter of other human beings.

'Witness' might just be the keyword here. Dictators kill from power, and from excessive fear and anxiety.

---

148 Catherine Heard, Exploring global disparities in criminal sentencing, Institute for Crime & Justice Policy Research at Birkbeck, University of London

149 Kyle Jaeger, How Prison Sentences in America Compare to Other Countries, Attn, January 20, 2017

> *"They see themselves as 'very special' people, deserving of admiration and, consequently, have difficulty empathizing with the feelings and needs of others ... Not only do dictators commonly show a 'pervasive pattern of grandiosity,' they also tend to behave with a vindictiveness often observed in narcissistic personality disorder."*
>
> *this is a topic that is perhaps the scope of this book. It comes back, however, two behaviors, and not just behaviors of the dictators. Awareness must come first, as many organizations are discovering about their equality and diversity training programs. Without awareness of human behavior, no change in behavior is really possible over the long term.* [150][151][152]

And lastly in my list of challenges that the world needs a solution to resolve: We allow a deadly virus to sweep across the globe, in spite of at least one (there are many, in fact) well-documented pandemic called the Spanish Influenza Pandemic of 1918.

And this again, perhaps, comes down to human behaviors. The tendency of governance is to throw money at a problem and hope it goes away. The tendency of governing authorities is to form a committee and do a study on the issue. And the tendency of individual citizens is to proclaim, "No! I'm not changing my habits because you say so!"

This is another topic that is certainly beyond the scope of this book, and lands nonetheless squarely in the public health policy box. But we've seen agreement around the world that public health policy must be a part of a sovereign state's responsibility to its citizens. If we are honest with ourselves, we have come out of COVID only realizing that we didn't do things right. On a national and global level, our leadership discovered how difficult it is to shut down an entire industry such as air travel, much less to shut down an entire country and its social, educational, and economic activities.

---

150 Seth Davin Norrholm, Samuel Hunley, The Psychology of Dictators: Power, Fear, and Anxiety, Anxiwety.org

151 Manfred F. R. Kets de Vries, Fighting Against Dictatorship, INSEAD, January 18, 2018

152 Maria Ressa, How to Stand Up to a Dictator: The Fight For Our Future, Harper, 2022

This and all the challenges I've named above give us food for thought, however, as we try to make a change through government-funded innovation in partnership with private organizations.

Yes, if we are pessimistic, we have a big mess on our hands! But luckily, we are having this conversation under the gentle watching eye of Dr. Martin Seligman, the doctor of positive psychology. We must stay optimistic and recognize that a big mess brings big opportunities. Besides, mess, complexity and chaos are part of life—as we have just learned.

When asked collectively and in public, world leaders state they want to work on these issues. But when asked what people are personally prepared to do about it, the answers become less focused. "Love, sweet love for everyone", but how?

Sociologists and behaviorists have long known that changing human behavior is a change that comes from within each individual; 'change by decree' or imposed behavioral change is not long-lasting (or even sees light of day in the first place) until the individual accepts change.

The Tools We Have

Let me take you back to some of the tools we have as we face the world's burning issues together. I'll use an example of the WEF and the climate crisis.

The World Economic Forum (WEF) holds its annual meeting in Davos, Switzerland. Every year it brings together thousands of business leaders, policy makers, non-governmental organizations, lobbyists, students, and others who want to ensure we are steering the world in the right direction. The theme for 2022 was "History at a Turning Point: Government Policies and Business Strategies." Discussion themes included health, climate, nature, business, technology and innovation, jobs, society and equity, and international collaboration. The Amazon rainforest, biodiversity, carbon sequestration education, technology innovation. Overfishing, clean water, energy, ethics, role of women, migration. Democratization, health, convergence of information technologies, war, organized crime, and cybersecurity. The thousands of people present were discussing dozens of topics that directly impact the future for all of us. The list of issues is long and multifaceted.[153]

I spent dozens of hours listening to these conversations online in order to fully understand the most severe and pressing issues we share, most notably the climate crisis and the speed of economic growth we expect.

The Intergovernmental Panel on Climate Change (IPCC) is a body of the United Nations responsible for advancing knowledge on human-induced climate change. IPCC has been warning us for some time that the type of growth we've seen over the past hundred years is not an option going forward. As economic developers, we have both the responsibility and the opportunity to rethink how we plan, carry out, and measure the economy.

I've been using the term "economic growth" and "prosperity" as synonyms. Have I been wrong in doing so? Does economic growth necessarily mean the growing use of natural resources—and even if it doesn't, is economic growth needed to build a meaningful, happy, healthy life for humans while restoring nature at the same time?

Post-growth economy[154] is a view, a proposal, a strategy, possibly a necessity, where societies operate without the push for constant economic growth.

There is also a group of economists who advocate for a slow-growth economy, arguing that it could be a sign of prosperity in that you have already reached the ceiling of your potential economic success, hence slower growth is the natural future.

You may have moved from primary production and manufacturing to a service-based, possibly a digital society. A slow-growth economy is also good against inflation. But this is very different from post-growth. Post-growth thinking includes the idea that instead of focusing on eternal growth which has led to assets and resources being accumulated by few. A post-growth society brings a focus on economic justice, social well-being and ecological regeneration to the forefront, where money circulates through the economy and reaches everyone. Here's a description from postgrowth.org:

153 The Millennium Project
154 The Post Growth Institute

> *Just as unlimited population expansion is untenable, so is unlimited GDP growth. Yet the open-ended commitment to economic growth persists, and it is now creating more problems than it is solving. It undermines jobs, communities, the environment, a sense of place and continuity, and even mental health. It fuels a ruthless international search for energy and other resources, and it rests on a consumerism that is manufactured by marketers and failing to meet the deepest human needs.*
>
> *Soon, developed countries will begin the move to a postgrowth world where working life, the natural environment, communities, and the public sector will no longer be sacrificed for the sake of mere GDP growth, and where the illusory promise of ever more expansion will no longer provide an excuse for ignoring compelling social needs. A postgrowth society will involve less consumerism and higher prices; quality of life will improve in ways too long neglected.*

Admittedly this sounds very optimistic—even borderline naïve—but we should not dismiss ideas just because we think they are not 100% achievable (much less enforceable) *today*. It is traditionally agreed that a healthy GDP growth for a country is between 2% and 3%, and many countries strive towards even higher growths. However, if you spread just a 2% GDP growth over 100 years there will not be enough rocks, natural materials, or places left for humans to live. This is a fact.

We are going to need to go from growth (at any speed) to regeneration (which must be consistent, planned, committed). This is what will have a deep effect on the economy.

The diagram below describes the flow of natural resources through the economy. Resources like water and minerals are procured from nature, and are used as input to the economy. Resources flow through the economy, are transformed into goods, and finally end up as waste and pollution. All of nature is powered by solar, but only a teeny-tiny fraction of this power is being harvested to run this economic operation. Instead, we use non-renewable energy that feeds into the pollution. We also do not harvest the thermal energy that our biosphere produces, missing out on an opportunity on that end.

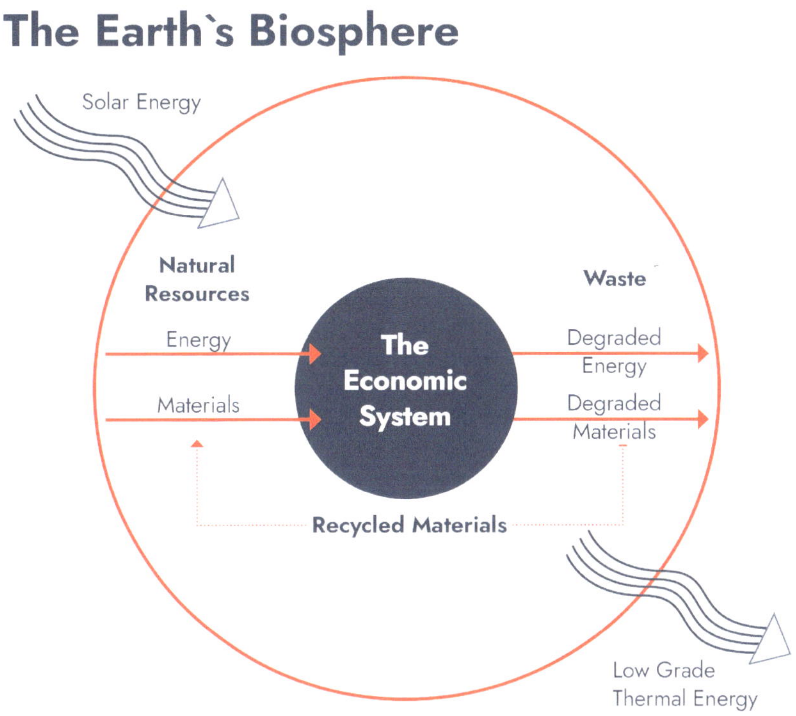

The Earth's biosphere as it is today, and what the ideal could look like.
Image credit Petra Soderling.

Energy and materials flow through the economic system creating waste

It is clear we cannot continue in this manner.

Gross Domestic Product (GDP), which measures the market value of all the final goods and services produced and sold, is an outdated model for tracking the health of an economy. It does not include the wellbeing of humans and nature. It only tracks the central square of our picture, but not the circle. The problem is that we live in the circle and not the square! We need new metrics that are both comprehensive and compassionate.

The Reality We Face

The people organizing and attending the World Economic Forum's events believe that the future is not just one economic paradigm, but many. I believe that is right.

Your future will depend on whether you live in a desert, island or forest, and on what type of industry skills you—these define what type of economic community you will have. As the postgrowth.org people predict, communities and their environment will have to put less emphasis on GDP, and instead focus on building economic wellbeing. Economic wellbeing will be born in communities and it will serve communities. It will also look different from one community to another—"many economic paradigms" make that a given.

The Covid pandemic launched a trend (that is still developing and unfolding) where people want to move away from cities into healthier, slower and lower-cost lives. Young people are saying "no thank you" to stress, cardiovascular diseases and obesity right along with jobs that hold no meaning to them, high rents and the high monetary and health costs of commuting. People are stating to pollsters that they would rather quit their jobs than go back to commuting to offices. The current employee market in the US is still resonating to this rude awakening as employers scramble to figure out exactly what to offer to candidates that will turn them into employees while still operating their businesses in an optimal manner. Businesses are changing their working conditions to adapt to these trends. Office buildings are being converted to housing units.

Digitalization is supporting this shift, with more physical and satellite broadband being built to cover even the remotest areas. Online tools, including augmented reality, virtual reality, even fragments of metaverse, are getting more usable, more affordable, and more available. Blockchain based digital identities allow people to identify themselves reliably for financial transactions, voting, and working in confidential government or commercial projects. New, cleaner technologies are bringing revenue and profits and creating jobs just like the old ones did. We saw that from the 1960s all the way through the 90s with the rise of the internet and the rapid deployment of all it potential that people were neither massively nor permanently put out of work due to one or more industries that disappeared, but instead found work in the new industry by learning the new tools.

This is the future of economic development.

What the world needs now is not just a litany of complaints, not just hand-wringing, about our shared problems. We are in need of an intellectual revolution where CEOs and policy makers, influencers, disruptors—all types of "movers and shakers"—agree to innovate past the 20th century business models where any change is viewed as a threat to the bottom line or number of jobs.

We have the tools. We have the brain power. We have the ability to communicate with each other around the world and share thoughts and solutions. The question really is: Do we have the collective will to use this capability for real change?

I circle back, always, to government and innovation. What is common to all of these challenges is the role of government. Governments everywhere have a role to identify and acknowledge the issues, certainly. But their job is also to address or resolve the issues identified, inspire people to take action and invite innovative solutions.

Governments have the authority to mandate and sanction, but as we are now wearing the King's crown from Robert Greene's 34th law of power, we choose to take the inclusive community approach. Talent is evenly distributed; opportunity is not. There are people all over the world—in your country and in mine, in our neighbors' countries—who have real potential to solve real, huge problems. We have a responsibility to ensure our children have a livable planet and a country where they feel safe, can prosper, and can support their own children in pursuing their own prosperity.

KEY TAKEAWAYS:

1. A limited planet makes it impossible to reach for limitless growth.
2. New methods and metrics, such as presented in slow growth, post-growth, alternative growth approaches, are imperative for the economy.
3. Economic development is the key to define what aspiration, growth, and success looks like in a limited world.

FIGHT CLIMATE CHANGE

Growing up, my parents and their friends were very conscious about the environment, the society, about everything in their world basically. They belong to the boomer generation which, as established, has been blamed for a lot of the things that went wrong, but who, in their youth, also did challenge the way things were going and the way they were being done at that time. Many evenings us kids would fall asleep listening to grown-ups make the world a better place around the kitchen table, while listening to Bob Dylan and local folk songs and gulping down cheap red wine. And smoking, of course. Lots of smoking!

The Club of Rome report "The Limits to Growth" came out in 1972.[155] Here's how they summarized the findings:

> *The earth's interlocking resources—the global system of nature in which we all live—probably cannot support present rates of economic and population growth much beyond the year 2100, if that long, even with advanced technology. In the summer of 1970, an international team of researchers at the Massachusetts Institute of Technology began a study of the implications of continued worldwide growth. They examined the five basic factors that determine and, in their interactions, ultimately limit growth*

155 The Club of Rome Publications

> *on this planet: population increase, agricultural production, non-renewable resource depletion, industrial output, and pollution generation. The MIT team fed data on these five factors into a global computer model and then tested the behavior of the model under several sets of assumptions to determine alternative patterns for mankind's future. The Limits to Growth is the nontechnical report of their findings. The book contains a message of hope, as well: Man can create a society in which he can live indefinitely on earth if he imposes limits on himself and his production of material goods to achieve a state of global equilibrium with population and production in carefully selected balance.*

As a child you don't really understand everything the grown-ups speak about, but you pick up the tone of voice (agitated) and some key words (tigers, DDT). I certainly learned about the Club of Rome. In my head it was a group of sweaty grown-ups with rolled-up sleeves, pen behind their ear inside a small room in Rome filled with cigarette smoke and discussing topics heatedly while typing reports on their typewriters. Maybe I wasn't off that much!

If the MIT computer in 1970 produced these results, imagine how accurate the newer models, calculated on today's supercomputers, would look. The Club of Rome has released over 60 similar thorough, well-researched publications about environmental issues nearly every year since that first one.

The United Nations first included sustainability in its programs in 1988 by establishing The Intergovernmental Panel on Climate Change (IPCC). IPCC works hand-in-hand with the World Meteorological Organization (WMO). IPCC works "to prepare a comprehensive review and recommendations with respect to the state of knowledge of the science of climate change; the social and economic impact of climate change, and potential response strategies and elements for inclusion in a possible future international convention on climate." The IPCC is the most comprehensive group of scientists that has come together to agree on the issue and the urgency.

Despite numerous studies, publications, and Nobel prizes, the scientists' warnings were largely ignored. It wasn't until Al Gore released a book, an Oscar winning documentary, and his lecture series 'the Inconvenient Truth' in 2006 that the time was ripe for wider masses to embrace the core idea spoken of for decades: that the speed and direction that humanity has engaged in is not sustainable. His graphics about the warming of the planet, the amount of CO^2 in the atmosphere, and animated maps about cities and states that would be left underwater hit home with many. Slowly a new generation of environmental activists, the Millennials, started to be vocal.

Politicians respond to what voters consider important. As more voters start to raise the environment as a top issue, more political action ensues. On Earth Day 2016, on April 22nd that year, the world was able to come together under the so-called Paris Agreement to agree how to mitigate, adapt, and finance measures to combat climate change. 196 countries signed it and 193 countries have since ratified it. The long-term temperature goal with the Paris Agreement is to restrict the increase to less than two degrees Celsius or 3.6 degrees Fahrenheit. In order to achieve this, emissions must be reduced as much as possible—50% by 2030—and be net-zero by 2050.

Source UN

Sustainable Development Goals (SDG) by the UN

The nice thing about the Paris Agreement is that each of the 193 ratifying countries need to report their progress to the United Nations. Voters all around the world can see what actions are being taken, and more importantly what results each country is achieving. The United Nations Sustainable Development Goals (SDGs) are 17 interlinked goals that measure overall well-being, hunger, health, equality, pollution, water quality, education, innovation, infrastructure, and so on. Each goal is broken down to more specific targets, color-coded and categorized. These targets are formed from over 230 indicators which were put together by the UN members and affiliated non-governmental organizations. This group is called the Inter-agency and Expert Group on SDG Indicators.[156]

There is no UN jail if countries don't follow through with their promise, but there is a name-and-shame list which hopefully will resonate with the voters in each country. In 2022, the top ten performers were Finland, Denmark, Sweden, Norway, Austria, Germany, France, Switzerland, Ireland, and Estonia.[157] Two of the world's wealthiest countries can be found further down on the list, the United States at #41 and China at #56, joining countries like Russia (#45), Israel (#49), and Brazil (#53). The poorest Sustainable Development Goals performers are the poorest countries in the world. Perhaps this is an opportunity for richer countries to demonstrate a broader perspective.

As a leader of a country, you will be happy to know that since the 17 goals cover such a broad range of topics, you don't need to only improve on environmental issues in order to get your ranking up the list. By investing in education, health, and innovation you are improving your overall score. For example, Finland being number one in the world at the time of writing this actually went down on goal number SDG13 which is 'Climate action'. Because of their good scores with education (SDG4), poverty (SDG1), clean energy (SDG7), and economic growth (SDG8) went up, their overall ranking went up.

---

156 IAEG-SDGs, United Nations Statistics Division

157 Rankings: The overall performance of all 193 UN Member States, The Sustainable Development Report

How your country scores with the SDGs is not a trivial matter. Financial institutions are in on the game and are making an impact on several levels. The World Bank, which provides loans, interest-free credits, and grants to developing nations, is a part of the Expert Group which tracks SDG indicators. They help the group with data and analysis on how countries are performing against targets that the World Bank covers. At the same time, the World Bank looks at the types of projects they fund in order to ensure each project works towards the SDGs. Finally, the World Bank as other major institutions have their own internal commitments to the 17 sustainability goals.

Major private banks are doing the same. Citigroup, JP Morgan and Chase, Santander, Bank of America, SEB, HSBC, BNP Paribas, Barclays, and other big funders have released their own commitments to the SDGs. In 2019, leading banks and the United Nations launched the Principles for Responsible Banking, with 130 banks collectively holding US$47 trillion in assets committed to climate action and sustainability.[158] This was one-third of the global banking sector, which means it is getting increasingly difficult to find market-priced funding for projects that work against the climate. Another acronym here—as if we didn't have enough already—is the ESG score. That's the Environmental, Social, and Governance score that lenders give to companies looking for a loan. The lower your ESG score, the higher your interest rate. This affects countries as well since S&P, Fitch, Moody's and others who issue credit ratings for nations and other regions, factor ESG into their calculations.

Despite all the positive developments, many still feel that not enough is being done to save the planet from destruction. I was watching a newsreel with a correspondent on German streets during the 2021 election for their new Chancellor. In this country that has been a trailblazer for recycling, renewable energy and reduction of pollution for decades, many voters felt the government had not done enough. Voters wanted stricter action. Generation Z led by thought leaders such as Greta Thunberg, Malala Yousafzai, or in the USA Amanda Gorman, Emma Gonzales,

158 130 banks holding USD 47 trillion in assets commit to climate action and sustainability, UN Environment Programme, September 22, 2019

David Hogg and Tyah-Amoy Roberts are voicing their generational determination to achieve protection for the environment, safer gun laws, and equal rights for women. All of these items are listed in the UN Sustainable Development Goals. Together with Millennials, this newest generation is a strong force not just as voters but as consumers. Politicians and businesses are taking notice.

The US government is catching up with the rest of the world with the 2021 and 2022 bills for energy infrastructure improvements. These new laws are designed to encourage the private industry to prioritize climate technologies. The US hopes to distinguish itself as a global green leader—a position that Europe has held for some time—by better integrating existing innovations into supply chains. If these incentives work as intended, more investment in innovation and climate technology deployment will create a greater production capability which could result in lower costs. New semiconductor and electric vehicle battery factories in American states are a good start. The world is rooting for America to pull this off.

KEY TAKEAWAYS:

1. The limits of our planet to support our lifestyle have been known for over 50 years.
2. Nothing has been done.
3. We are now experiencing the results of this inaction, and finally waking up.

FIGHT FOR DEMOCRACY

In December 2016, my husband and I were watching the US presidential election vote counting on TV in our home in New Orleans. We had bought champagne to celebrate the country's first female president. Hillary Clinton of 2016 represented the type of politician who has devoted her life to working tirelessly to make the world a better place. She's done so by using political methods, running for elective office, passing legislation and working through established guidelines. I respect that approach, as much as I respect the Hillary Clinton[159] of the 1960s who was organizing, marching, and creating grass root action.

When it became clear that Donald Trump had won the presidency, I was flabbergasted. How can it be? He received fewer votes. It did not seem possible that in the United States of America, the oldest democracy in the world, the candidate with fewer votes could win an election (and I send you back to the one-man-one-vote commentary I made earlier, as the US Electoral College fits that history). As it turns out, we were approaching a time when the ideal of a peaceful and democratic election of politicians who would server their countries was being challenged. It was clear this president was not here to serve the country, but for much more narrow and short-term goals.

In Brazil, Jair Bolsonaro ran a Trump-style election campaign in 2018 and was inaugurated as president of Brazil in January 2019. His campaign

159 About Hillary, The Office of Hillary Rodham Clinton

messages included harder conditions for women to work or do family planning. He wanted to reduce the rights of minority groups including sexual and religious minorities. He encouraged physical violence as part of a normal political discourse.

On the other side of the world in Hungary, Victor Orbán has presided as prime minister since 2010. He has been democratically elected, but has pursued policies that reduce democracy in his country. These include limiting the freedom of press, erosion of the judicial system's independence, and various ways to keep competing parties out of power. Orbán and his right-wing party have been critical about the European Union while at the same time accepting its money and redistributing it to his allies and family. During his time in office, Hungary has dropped significantly in both the Press Freedom Index[160] and the Democracy Index[161].

In nearby Poland, the populist party Law and Justice gained power in 2015 and passed measures that increased political influence over state institutions and threatened to reverse Poland's democratic progress. The Polish people have seen an increase in nationalist and homophobic rhetoric while supporting the "traditional" Catholic Polish lifestyle. Some East European nations are following suit.

Other countries sliding down the democracy index are China, Brazil, Chile, Ecuador, Paraguay, Mexico, Tunisia, Kyrgyz Republic, Haiti, Lebanon and Spain. Only 6.4% of the world's population live in "full democracies", and the overall democracy score in the world is now lower than it has ever been since it was first tracked in 2006.[162]

While it is true that democracy (and most election processes) has a built-in ability for voters to challenge their leadership, we need to ask what the elected leadership is doing in relation to voter expectations. Erosion of democratic principles is not fiction.

---

160 RSF's 2022 World Press Freedom Index : a new era of polarisation, Reporters Without Borders (RSF), 03.05.2022

161 Democracy Index 2021: the China challenge, Economist Intelligence Unit

162 Democracy Index 2021: less than half the world lives in a democracy, Economist Intelligence Unit, February 10, 2022

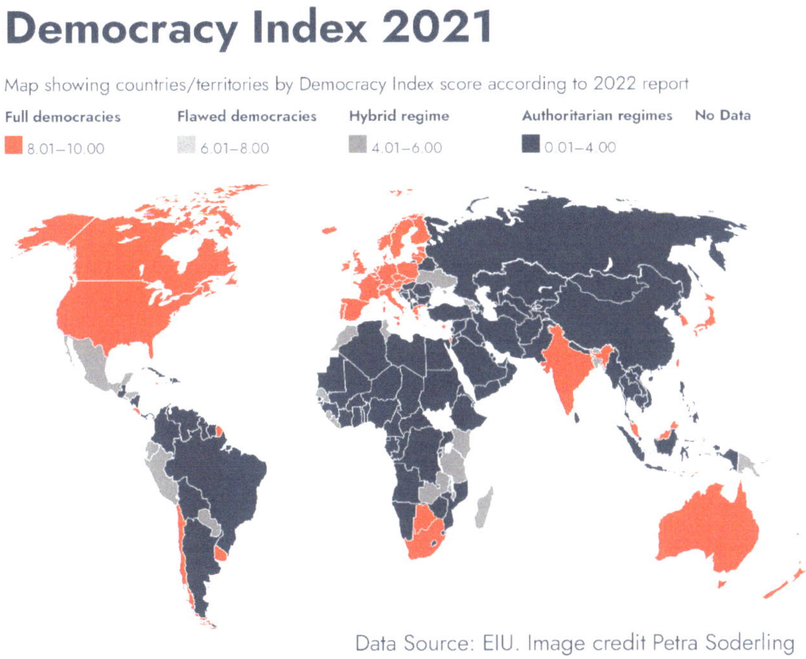

Less than half of the world lives in a democracy

Democracy is good for business. You may not be surprised if I again list the same countries as most democratic, as I have listed them at the top of most other lists throughout this book measuring wealth, progress, and overall satisfaction. The top five most democratic countries are Norway, New Zealand, Finland, Sweden, and Iceland. The top five countries from the Press Freedom Index are Norway, Sweden, Denmark, Estonia, and Finland. As a reminder, these are also the world's most innovative nations.

The lowest-ranked countries in these two lists are also performing worst in financial metrics. As authoritarian Russia attacked Ukraine in February of 2022, the global community quickly raised sanctions against Russian companies and individuals. As a result, international companies left the country and governments reduced procurement of Russian energy. The Russian economy has suffered badly.

Speaking of the freedom of the press, the one thing all these authoritarian leaders have in common is using misinformation and disinformation to confuse people, allowing their own agenda to thrive in both traditional as well as social media. Their strategy is working as long as a good number of people in social media do not have adequate "media literacy" skills. The Council of Europe—an international organization founded in the wake of World War II to uphold human rights, democracy and the rule of law in Europe—has addressed media literacy in Europe for over two decades. They describe it as follows[163]:

> *Media and information literacy is the main tools for empowering people, communities, and nations to participate in and contribute to global knowledge societies. In the view of the Council of Europe, it is of utmost importance for individuals to be able to develop cognitive, technical and social skills and capacities that enable them to effectively access and critically analyse media content; to make informed decisions about which media they use and how to use them; to understand the ethical implications of media and new technologies, and to communicate effectively, including by creating content.*
>
> *Media and Information Literacy (MIL) brings together three distinct dimensions: Information Literacy, Media Literacy, along with ICT* and Digital Literacy.*
>
> *Information and Communication Technologies (ICT)

The Council of Europe has two objectives for media and information literacy. The first objective is to set Media and Information Literacy within institutional frameworks. The second objective is to provide tools for empowering individual media users of all ages and walks of life.

In the United States, media literacy issues surfaced in the wake of the 2016 presidential election, and the ensuing investigation on Russia's involvement in American politics by special counsel Robert Mueller, as appointed by Congress. Mueller's team found that Russian interference in the 2016 presidential election was illegal and occurred "in sweeping and systematic fashion." Congress has since interviewed social media

163 Media Literacy, The Council of Europe

company CEOs about their role in the spread of misinformation, but no political action has been taken as of 2023.

Grassroot movements, however, are taking media literacy seriously, and several NGOs are collecting data and educating the American people on how to address and consume content in social and traditional media. Media Literacy Now is a nonprofit tasked with educating school children about the issue. In their latest policy report from 2021, they claim that only 14 of the 50 states have a statewide education policy for media literacy or digital citizenship and this number had not improved since 2020. Media Literacy Now is working with the Europeans to rank countries for media literacy and their latest press release states that "A new index shows that the US scores low on media literacy education".[164] As the USA is still the leader of the free world, this is an alarming piece of information for democracy.

There are major systemic issues in all countries that work against democratic processes. Even in countries that score high in Transparency International's Corruption Perceptions Index[165], there is corruption. It may not be transparent, or take on the form we typically associate with corruption, but corruption is present. Every day, an army of industry-paid lobbyists are meeting with politicians and civil servants helping them write and argue for budgets, legislation, and other public measures. This is in itself a useful function; how else would lawmakers know about the weeds of let's say, telecommunications networks? Having been on the industry side explaining the importance of certain technology advancements to lawmakers, I don't condemn the relationship. It just needs to be transparent enough so that bribery, a crime of opportunity (a real form of corruption), is hard to carry out.

Just as voters haven't been educated to spot fake news, they also haven't been educated to read and interpret political discourse. Increased partisanship and polarization, lack of political education, and manipulation of public opinion are spiraling countries even further from any

164 Joe Carr, A new index shows that the US scores low on media literacy education, Media Literacy Now

165 Corruption Perceptions Index 2021, Transparency International

open debate of the issues. Non-political issues are becoming politicized; political issues are being divided by party lines. Add to this the work-from-home trend—as people move away from company offices and into home offices, and from cities into areas where they connect to social circles mainly through social media—there is a risk of isolating worldviews that morph deeper into a bubble of like-minded friends. These bubbles can be vulnerable to political or commercial influence.

It is not just voting-aged individuals but younger people who are fighting for democracy along a number of lines. School shootings in the USA are forcing children and their parents to lead campaigns aimed at policy makers on gun safety. 17-year-old Malala Yousafzai, who was shot in the face by Pakistani Taliban for speaking out for girls' right to go to school, has risen as a global leader in speaking out for women's rights. It was in the UK that students were first to realize that the police and private companies had been rolling out and using facial recognition AI technology around the country, taking snapshots of people's faces without their consent. The images were used for various purposes from marketing campaigns to criminal investigations; wrongful arrests of many and, notably, people of color took place, as the technology did not work 100%. It was a group of young people started the Big Brother Watch[166] with volunteers around the country who started collecting data and cases to put pressure on lawmakers.

Despite young citizens taking action across many issues, the future of democracy is not safe unless countries take firm action. A 2017 study showed that half of Millennials struggle to identify false or fake information online and that half of Millennials also receive and share their news through social media.[167] Yet, in the Deloitte Global 2022 Gen Z and Millennial Survey from 46 countries, Gen Z's top five concerns were cost of living, climate change, unemployment, mental health, and sexual harassment. For Millennials, the top five concerns were cost of living, climate change, health care and disease prevention, unemployment, and

---

166 Stop Facial Recognition, Big Brother Watch

167 Bulldog Reporter, Fake news epidemic: Half of millennials struggle with news literacy, Agility PR Solutions, May 16, 2017

crime and personal safety. Just because young people aren't identifying the role of democracy (or media literacy) as a top concern doesn't mean that democracy or lack thereof would not affect their lives in a dramatic way.

It is our responsibility to be the adults in the room. Whatever our age, we are part of a generation that has this power. We need to be taking care of the planet and the future generations, and make sure they see us. Too many times, I've spoken with young people who don't vote, don't care, and don't bother because "It's not gonna change anything." If young people don't vote, the bad guys win. If young people don't trust the system enough to run for office themselves, the bad guys win. If Millennials and GenZers worry about cost of living, climate change, health care and sexual safety—then as law makers, civil servants, journalists, and economic developers, we should all be working to fix these issues so young people can feel like they belong here.

Everyone plays a role in ensuring democracy wins, even those who think, "It's not gonna change anything." Inaction is also action, in this case action against democracy. Those who actively allow fake news and hate speech to flourish are just as complicit as those who silently accept it. Those who vote for (or re-elect) a candidate who promised good for all but acts in a divisive and hateful way once in office are complicit. Those who plan and execute commercial actions that destroy nature are complicit.

Let's take the high road and help young people understand what a wonderful system democracy is.

KEY TAKEAWAYS:

1. Democracy requires awareness, literacy, and the will of all citizens to protect it in order to create prosperity and the good of all.
2. Expansion of democracy is not self-evident, in fact counter-democratic forces have strengthened.
3. Each country separately, and all countries collectively, need to address the systemic flaws that threaten democracy.

FIGHT FOR GLOBAL HEALTH

Throughout history, economies have risen and fallen with famines and pandemics. The Cocoliztli epidemics in Americas the 1500s, Indian famine in the 1600s, the Indian and Chinese famines in the 1700s and 1800s, the post-WWI Spanish Influenza, and HIV/AIDS in the later 20th century each killed large populations and seriously harmed both social and commercial activities.

Economic depressions tend to create friction between groups of a population, and lead to violent conflicts, even war. Together with pandemics, famines and subsequent wars, we've seen global wealth follow the trend in the form of great depressions and even the collapse of entire empires.

In 2020, the world shut down due to the Covid-19 pandemic. We are still struggling, as I write this in early 2023, to figure out how to live with it. Are we stuck with this virus forever? As we've seen, a pandemic is not just about individual or public health. It's equally about the economy, social structures and the climate. For the first time, 2020 had cleaner air and cleaner waters than many years prior. There was no traffic on the streets; people took up biking and walking again. Arguing for healthier citizens is arguing for a cleaner environment and a functioning economy.

Even though many countries had a plan in place for health events like the Covid-19, every government struggled with how to manage it when it was upon us. The need for decision-making abounded: Which institutions to close and when and for how long; whether a mask mandate is required; where and how to get vaccines and hospital equipment; how to manage essential and other supply chains across a nation? Overall, how to manage this unusual situation legally?

These were and still are hard questions that have multiple answers.

I recently discussed their respective Covid-19 experiences with a group of European twenty-somethings. They had a very critical view on how their countries—Finland, North Macedonia, and Spain—had dealt with the pandemic. These young people felt their lives were unnecessarily put on hold. They had not been able to go to school, see friends, go to the movies or concerts for nearly two years. Being against the vaccine, they felt that their society was pushing for decisions harming their health and forming a two-tiered system where the vaccinated have rights and the unvaccinated become pariahs. This group would rather have seen that the vulnerable population was isolated from the healthy and invulnerable ones, instead of closing everything for everybody. No doubt many countries, states, and cities were considering this option but chose otherwise.

In Colorado, I spoke to a neighbor who is a nurse in a Catholic nursing home in Denver, and who worked the night shift throughout the pandemic. Her view was different. She described her experience to me:

> *At first no one knew what this virus was, other than it came from China. We were all in an unsure environment, but you cannot stop caring for the elderly just because there was no science or vaccine available. Everyone in healthcare was lost in those early days. Many left, but many very brave people also stayed. For example, there was a nurse in another location, a Denver hospital's Intensive Care Unit (ICU). She was scared but she was forced to go into the ICU unprepared. She caught Covid and passed it on to her family. She quit her job in the hospital and came to work in our nursing home where she felt safer. You had to choose between sacrificing your career, or your life, or the life of your family.*

> *I am used to death; this is a nursing home and people die, but this was different. We assigned a certified nurse specialist who was keeping up-to-date with the CDC guidelines and FDA approval schedules. It's made us all feel safer, and luckily in our nursing home no one died of Covid. We did have one certified nurse assistant who didn't feel safe taking the vaccine, but also felt uneasy going to the patients' rooms unvaccinated so she left. It was the fear of the unknown: an unknown vaccine and an unknown disease.*
>
> *It was also very hard for me. I am not a young nurse. We had nurses who were pregnant, so I went into the rooms first to protect the pregnant ones. We are the last resort for our patients. Us nurses must be responsible on the floor and respect the individual. I had 20 residents on my floor in my night shifts, plus eight nuns who stayed throughout the entire Covid-19 epidemic praying and caring for the elderly.*

My neighbor never thought she would change her career during or after Covid. She thinks the US government did an excellent job protecting its people. They invested in the vaccine, relied on science, and communicated everything clearly.

At the time of writing this the world has seen 561 million Covid-19 cases, and 6.37 million have died. Compared with the Spanish Influenza of 1918 with its 50 million deaths, this is not a lot of deaths. But in our lifetime it is unprecedented. We can only imagine what the death toll would look like without the drastic government measures that were taken—closing schools and restaurants, restricting travel, and mandating masks and vaccines—and that put our young friends in a state of distress.

A pandemic is not just a health tragedy, but also leads to serious economic failures. The International Monetary Fund IMF estimated in January 2022 that the total cost of the Covid-19 pandemic will rise beyond $12.5 trillion.[168] To put that in context, the United States' GDP in 2020 was $20.94 trillion. Who knows what the final cost will end up being when looked at, say, a decade from now?

---

168 Macro Matters, IMF sees cost of COVID pandemic rising beyond $12.5 trillion estimate, Reuters, January 20, 2022

The IMF calculation of $12.5 trillion factors in pandemic-induced costs from supply chain disruptions, inflation, and tighter monetary policies. Their estimate anticipated that, going forward, huge gaps in Covid-19 vaccine rates, a widening divergence between rich and poor countries, and friction within countries caused by the pandemic—along with learning losses and increased negative gender impacts—would cause more protests, tensions and insecurity. All of this will feed into poorer economic outcomes than if we'd never experienced a health pandemic.

You may be thinking, "Well, this is all speculative. We don't have a parallel universe where the pandemic didn't happen, so we cannot really say for sure which economic glitches are due to the pandemic and which are just seasonal fluctuations that might've occurred anyway." If you're only interested in the direct cost of addressing Covid-19 as an illness, there are numbers for that, too. There is a cost calculation framework that was done for previous pandemics that we can use.

The Global Fund[169] is a fund that was originally founded to Fight AIDS, tuberculosis (TB), and Malaria. In the beginning of the pandemic, they pivoted to addressing Covid-19. This group has the tools, procedures and experience in how and where to invest to support countries as they fight a deadly epidemic.

In order to discover the cost for pandemic preparedness, the Global Fund took action-based cost estimates generated from prior research that estimated how much it cost to build sufficient country-level capacity for health security against deadly infectious diseases. They then coded them as costs that are directly or indirectly related to these health security building efforts, and costs that are unrelated to them.

Activities and costs directly related to deadly infectious diseases included skilled healthcare workforce trained in surveillance and infection prevention and control, select rapid diagnostic tests, and point of care diagnostics. Support for national-level supply chain capacity, including warehouse space to send, receive, and maintain an inventory

169 Stephanie Eaneff, Matthew R Boyce, Ellie Graeden, David W. Lowrance, Mackenzie Moore and Rebecca Katz, Financing global health security: estimating the costs of pandemic preparedness in Global Fund eligible countries, medRxiv, March 07, 2022

of medical countermeasures is also a direct cost. You may remember how critical the supply chain was in the beginning of the pandemic, with most countries scraping together protective gear, ventilators and other hospital equipment.

Activities indirectly related to how the Global Fund has addressed deadly infectious diseases included procuring laboratory equipment and consumable materials associated with virus culture, serology, and Polymerase Chain Reaction (PCR) capabilities. It is important that countries have capabilities to study and identify viruses to create vaccines and therapeutics. Stipends and supplies for annual cohorts of full-time Field Epidemiology Training Programs (FETP) are also activities with indirectly related costs. The support for national-level risk communication specialists and broadcast time for proactive health communication to the wider public is likewise a relevant cost.

Activities and interventions from prior research that did not apply to the Global Fund's framework and were left out from the final cost analysis included personnel and supplies to respond to chemical and/or radiological emergencies, and physical infrastructure for Emergency Operations Center (EOC).

To pull this all together, just the cost for tracking, diagnosing, and treating Covid-19 costs $124 billion globally. That's the GDP of a mid-sized country or annual revenue of one of the largest companies in the world. Add the cost of sub-performing economies, that's the $12.5 trillion estimated by the IMF. It quickly becomes apparent the world needs innovations, solutions, decisions, and practices to ensure we wouldn't end up in this situation again.

Unfortunately, though, we almost certainly will. The Covid pandemic, with its mask and vaccine mandates and other government interventions are one of the categories of flammable misinformation going around the world in social and traditional media. The efforts to prevent a health crisis include a determination to educate the public with proven facts and information in an environment of misinformation.

Covid-19 or other global epidemics are not the only health-related threats that we have. As we are all coming out of the pandemic, it is natural for us to think about this large-scale sweeping disease when it comes to health events. The top 10 deadliest diseases in the world, however, do not include Covid-19, but are heart disease or coronary artery disease, stroke, lower respiratory infections, chronic obstructive pulmonary disease, lung cancers, diabetes, Alzheimer's disease and other dementias, dehydration due to diarrhea, tuberculosis, and cirrhosis. That's a lot of disease. The sad part is many of these deadly diseases are preventable. In some cases, it takes education for hygiene and healthy lifestyle; in other cases it takes the deep pockets of government-funded research to find the cures. In both cases, problems are solvable if we so choose.

The Bill Gates and Melinda French Gates Foundation is well known for its work on improving global health, and it is working on these specific problems. The Foundation, with about $50 million on its balance sheet, operates by awarding grants to research and development in education, healthcare and diseases, agriculture, environment-related issues, and social issues. Human ingenuity and innovation are at the core of their operations in supporting solutions to the world's biggest challenges. Grants are awarded based on free applications, or as part of Grand Challenges. These Grand Challenges are initiatives fostering innovation to solve key global health and development problems. There are currently over 200 Grand Challenges open, a vast majority of which are solving health-related problems. Since 2003, the Foundation has awarded funds to more than 3,500 health and development projects in almost 120 countries, amounting to more than $6 billion in value.

The Bill Gates and Melinda French Gates Foundation is perhaps one of the world's brightest spotlights shining on specific problems in specific areas and fixing them one by one. Examples of the types of issues being tackled include these five currently open calls: 1) Interventions to Enhance Epidemic Intelligence, Surveillance, and Outbreak Response, 2) Strengthening Modeling and Analytics Capacity and Ecosystem for Women's Health, 3) Pathogen Genomic Surveillance and Immunology in Asia, 4) African Agriculture Climate Adaptation Research System, and 5) Strengthening the Contraceptive Research and Development Ecosystem in Africa: Accelerating Innovations in Non-Hormonal Contraception for

Women. Even though the amount of money awarded does not match The World Health Organization's $7+ billion annual budget, this philanthropic foundation is a critical component in global health innovation.

Health is more than the absence of disease, though. The World Health Organization (WHO) lists these additional, perhaps surprising, threats to global health: climate crisis including air pollution; health care delivery in conflict areas; unsafe products; low pay of health care workers; adolescent safety including suicides, HIV, road injuries; lack of public trust of health care workers; and unregulated technological advancements such as creating new organisms that can harm people. I'm adding these to what the Bill and Melinda French Gates Foundation calls 'must-solve problems.'

Much like climate and democracy, global health is not an isolated issue. These are massively complex problems with mazes of interdependencies across domains and the world. Problems on this scale are hard to solve, but try we must. Who knows, maybe a hundred years from now historians will look at our era and pinpoint the Covid-19 pandemic as the moment in history that led to unprecedented inflation and economic crisis, exacerbating misinformation online and the divisiveness of people, ultimately leading to the collapse of democracy? What they will see is left, perhaps, is an authoritarian rule in a world of one public health disaster after another. Or maybe, they will observe that people soon realized this is a undesirable outcome and started behaving in ways that supported their own health and the health of their communities.

KEY TAKEAWAYS:

1. Global pandemics are very expensive to tackle after the fact.
2. Having to make political and societal decisions that limit people's freedoms will have long-term negative consequences on trust.
3. There are global governmental and non-governmental organizations working on innovations that prevent and solve health issues.

YOUR HOMEGROWN ASSETS

Congratulations for making it this far. By now we have established that governments play a critical role in launching and enabling innovation. We have learned about the sets of tools at your disposal, and we have looked at examples from some countries that have been using these tools in order to spur innovation and economic development. We've discussed what the world needs and some of the ways to go about making a change. Now it's time to focus on you.

Whether you are a current or aspiring politician, a civil servant in a ministry or federal department, or you work for a local economic development agency or investment promotion agency—you are concerned about the economic well-being of your region. You are a voter and a taxpayer who wants to see your country thrive. You are perhaps a parent wishing the best future for your children and theirs.

In all of these roles and before making any change, you have to set a baseline to show your starting point. What does your region look like? Are you up in the mountains like me at the moment? Do you have bodies of fresh water? Are you on an island or perhaps in the arctic region? Maybe you are in a small town with a small population, like Palo Alto, CA was in 1950 with just over 25,000 people. You could be a small country with a large population, like Singapore which is a 733 km2 (283 sq mi) island where 5.6 million people reside. It doesn't matter what your city, state or country looks like. Places as different as Monaco and Australia are building success in their own ways.

Classical economics offer these three types of resources: Land, Labor, and Capital. Your landmass, vegetation, water, and minerals typically remain whatever they are. Unless you are planning to overtake your neighboring country, this is the "homegrown asset" set you have to work with. It is your starting point.

Land

Depending on history, the landmass you are starting with may be well cultivated or in a more natural state. Some countries have a long history since the Industrial Revolution of manufacturing goods and building infrastructure. Japan is a good example with its oldest company Kongo Gumi, a construction company, established in 578 AD. Since those times, Japan has had phenomenal success using the land they have in creating new industries and prosperity for its 125 million citizens living on its 421 habited islands. Japan is famous for inventing some of the cornerstone technologies and business strategies for the manufacturing industry in the 20th century.

Some countries are in a more natural state. The state of Hawaii, while still developed, has an abundance of beautiful nature which it is offering for tourists, marine and space scientists, and as a filming location. A quarter of the Hawaiian economy comes from tourism. With over 10 million tourists visiting the islands in 2019, tourism provided jobs for 216,000 locals and left $18 billion in the Hawaiian economy. Hawaii has a number of startup companies building technology solutions for the food, culture, travel, and environmental sectors, using its nature and culture as an inspiration, and the travelers as a marketing channel and user base.

Oil, gas, and minerals—part of the land you have—are not evenly distributed around the globe. Some countries and states have more luck than others. As the world moves away from fossil fuels, economies that have been relying purely on those sources are now facing renewal issues. Saudi Arabia is a case in point and is executing on their 2030 vision for diversifying its economy, using the monetary reserves it has accumulated over decades. Dubai and other United Arab Emirates have similar strategies, investing in startups and diverse technologies.

Minerals however are a different story. As we are moving towards electric vehicles and renewable energy, the need for batteries is exploding and so is the demand for the minerals needed to produce them. Minerals are also used in communications networks, computers, construction materials, and in agriculture. Countries with large mineral sources, such as China, Russia, the USA, Australia, Indonesia, Brazil, Chile, or South Africa are the potential winners in this game. But being economically successful based on natural resources alone is not a good end game. Looking at this list of countries, we can see some underperformers when it comes to using that revenue to grow its total economy.

The classical economists did not mention air, but I want to bring up the underestimated role of air quality to the economy. Poor outdoor air quality kills more people than wars, car accidents, or cigarette smoking. According to the American Psychology Association APA, high levels of air pollution may damage children's cognitive abilities, increase adults' risk of cognitive decline and possibly even contribute to depression. In fact, a study published in 2017 on the impact of sustained exposure to air pollution on life expectancy, China's Huai River Policy[170] is quite remarkable. This policy was put in place in the 1950s, and it dictated that in order to save on heating expenses, the Chinese government would divide the country in half, right along the east-west course of the Huai River. Residents north of the river received either free or subsidized heating coal, whereas residents south of the river did not. In later studies that looked at 40,000 people from just north and south banks of the river, researchers found that the people born to the side with free or subsidized coal had about three years lower life expectancy, their children completed one year less of school, and earned 13% less than their peers on the south side of the river. Air pollution from that coal had a massive impact on the people, educational attainment, and the local economy.

---

170 Avraham Ebenstein, Maoyong Fan, Michael Greenstone and Maigeng Zhou, New evidence on the impact of sustained exposure to air pollution on life expectancy from China's Huai River Policy, PNAS, September 11, 2017

Labor

Our second type of classical economic resource is the labor force you start with, also called human capital. Your tools for influencing the size and type of your labor force include wage and taxation decisions, education and immigration policies, and the availability and quality of healthcare services. Setting a minimum wage or wage categories where you can, and taxing different industries differently will create pull-and-push effects that you can use to support your strategy. Planning ahead for the curricula, number of teachers, and the number and locations of elementary schools will result in a certain type of labor force many years from now, while creating strategies only around higher education has a shorter-term effect. By investing in certain fields of study, for example natural sciences or technology, you could be responding to industry needs or simply creating a pool of talent that will itself push your economic strategy towards a chosen goal. Having an immigration policy that attracts certain types of students and workers affects your labor pool. Offering preventive healthcare and accessible treatment helps ensure that your labor force is not suffering from preventable and curable illnesses that might affect first, their ability to do sustained work, and second, your economy.

Whether your baseline is in manufacturing, services, or basic education, you have something to build on. You may have a small or large population, but they all go to school, work and interact in their own domains every day. People all over the world have skills in technology, engineering, construction, commerce, care, education and so on. These people want to develop their personal skills and they want the opportunity for their personal skills to contribute to the prosperity not just of their family but of their country.

Capital

By using your land and labor to create economic prosperity, you also ensure access to capital—money—that will allow you to invest further. By investing taxpayer money into developing your agriculture, forestry, infrastructure, defense, R&D, startup businesses, education, and healthcare, you are not only growing your economy, but sending a signal to lenders that you are willing to invest in your own future. Both of these are important factors that affect your access to and the price of capital.

The three largest private credit ranking agencies are Standard & Poor's, Fitch, and Moody's. They give a credit rate to each sovereign nation in the world, in the form of a letter grade. The better your grade, the more credit you have available, the longer your loans can extend, and the lower your interest rate is.

It doesn't matter whether you're a AAA+ or B- or even lower, there's always a way to fund development. There are institutional finance organizations that focus on funding developing countries so they can invest in advancing their economies. What is known as The World Bank is actually two separate funds set up for this purpose: The International Bank for Reconstruction and Development (IBRD), which lends to middle-income countries, and the International Finance Corporation (IFC) which helps the developing countries. Regardless of your economic and financial situation, there are tools available to develop your country further.

KEY TAKEAWAYS:

1. Your economic development assets are nature, labor, and capital.
2. The hand you were dealt—your starting point—does not determine your future outcomes.
3. You will always have access to capital.

NATURAL RESOURCES: LAND, SEA, AIR

Let's go back to the land you start with and give examples of how this works.

Canada, Finland and Russia have forests. Norway, the United States and Saudi Arabia have oil. Australia, Argentina and China have minerals. Coastal regions and islands have wind, fish, algae, and other marine resources. Every country or state has a unique natural resource that they can deploy.

I recently had an argument with a friend about this. My claim was that even if not immediately visible, there is a market for whatever each region has to offer. He was pushing back saying there are places where there's nothing at all to offer. He gave examples of states that I'm not going to name here, but my argument is that even if you have "a vast nothingness," you can use that space for something. You could create a regulatory sandbox for testing scientific experiments of self-driving cars, drones, or any other technology that needs to be tested far from people and infrastructure. If you have marshland, it could be an optimal location for a biological science center or maybe there is some kind of microbiological organism that you could harvest and refine into something sellable or otherwise valuable. Impermeable granite or shale rock located in non-volcanic areas are ideal places for storing nuclear waste.

My point to my friend was this: There's always something to be developed as long as you are on planet Earth. And as long as you are thinking innovatively. Why, there's a startup company that makes protein out of air!

The top ten countries based on the monetary value of their natural resources are in this order: Russia, USA, Saudi Arabia, Canada, Iran, China, Brazil, Australia, Iraq, and Venezuela.

Russia's natural resources in 2021 currency are valued at $75 trillion. Russia has crude oil, natural gas, coal, and rare earth metals. It also ranked third in the world in the production of industrial diamonds. But as we can see, having all of this wealth sitting underground, having assets doesn't mean you can use or are using them for your benefit. Russia has jeopardized their opportunity to be the wealthiest country in the world with catastrophically bad management.

Some people consider forests important because of the opportunities they provide to produce timber, a product that has been relied upon for shelter and other types of construction. Others understand the vital importance of forest lands for the health of the planet. Forests provide us with oxygen, remove harmful carbon dioxide, and help regulate temperatures. They also balance the ecosystem by nurturing and protecting countless species of flora and fauna.

Woods, water, mountains, wind, soil, minerals and gasses are natural resources that humans have deployed and exploited for hundreds if not thousands of years. We have done so in a way that has been harmful to the environment and to our own health. To be sustainable in economic development doesn't mean giving up these resources. It just means you can use them in a smarter, cleaner way.

And what if you don't have oil, forests, or minerals? As you can see from our top ten countries, just having natural resources doesn't automatically make you economically successful. In the same fashion, not having these resources does not make you automatically economically unsuccessful. It's all how you understand what you have, and then how you decide to innovate with it and manage it.

Wood has been used for centuries for building, heating, and more recently for making paper and paper products. These are quite low value-add uses, compared to the use of battery or computer making minerals, as an example. So some countries are investing in researching new uses of wood. Wood-based materials can now be turned into plastic and thus are replacing bags, computer chips, clothes, bathroom furniture, displays, car panels, body parts, tires, water bottles, and so on. The water and sewage market is expected to grow at a rate of 7.1%, from US$578.41 billion in 2021 to US$852.92 billion in 2026, by opening up new product development opportunities in digital water management, wastewater processing, filtration, flood prevention, water-saving technologies, decentralized infrastructure, materials, and desalination.

I mentioned the testbeds for unmanned aerial vehicles and autonomous vehicles. In the United States and within the European Union, testing and operating these types of gear is strictly regulated. We don't want random drones or driverless cars roaming around our streets, airports, historical sites or schools. There's a reason why usage is regulated. But developers of these devices need to test them in different weather conditions and different altitudes, so there is a demand for my friend's so-called "vast nothingness", too. There are other types of equipment that need testing, possibly in windy conditions. Electric vertical take-off and landing (eVTOL) aircrafts are helicopter-style drones that can carry people. This plane takes off vertically from any platform without noise, without pollution, so testers can use that wetland for testing takeoff from a swamp. It can fly hundreds of miles without a need for a road or track, and land on the top of the mountain, so testers need access to those kinds of spaces and places as well.

Other things to set up if you have "nothingness" could be space and weather stations. These are ground satellites and sensing stations that need to be in a particular angle towards the night sky, or cover the earth in its entirety. The World Meteorological Organization (WMO) has over 8000[171] ground weather stations running all over the world. It's highly likely there's one in your state or country. Instead of just accepting a modest rent for this land use, you could think of joint research or development projects around them that is also revenue-generating and jobs-creating.

My former home state of Louisiana has been able to turn a state that used to rely mainly on oil and gas and polluting chemical industries into

171 WMO Weather Stations, The World Meteorological Organization (WMO) February 22, 2018

a diverse state economy that enjoys Hollywood film productions, computer-generated images, 3D modeling, and other computer graphics-related industries. We have the Louisiana Economic Development (LED) and the Greater New Orleans (GNO), Inc to thank for this. They also invest in science. This bore fruit in 2017 when Louisiana State University (LSU) was one of the recipients of the 2017 Nobel Prize in Physics. The prize was awarded to the pioneering leaders of the Laser Interferometer Gravitational-wave Observatory (LIGO), for the first detection of gravitational waves. The LIGO system consists of two large observatories , one in Louisiana and the other in Washington, with the aim of detecting gravitational waves by laser interferometry. These observatories use mirrors spaced four kilometers apart which are capable of detecting a change of less than one ten-thousandth the charge diameter of a proton. Having your own physics research, and being part of these research communities is what brought Louisiana this opportunity.

As was the case with Hawaii, beautiful nature can be capitalized in a way that respects the natural state of your natural areas. Nature photography, individual health packages, and spiritual tourism are services that are highly lucrative because there is more and more in demand by the modern traveler. Airbnb recently launched a service they call Airbnb Experiences. This is a service where anyone in any location can offer a service to Airbnb guests. Airbnb has become not just a provider of overnight sleeping facilities but is more a facilitator of the whole hospitality package. According to Airbnb, travel is changing. Before people used to plan their travel around a destination. Let's say I want to go to Paris. The next step is to find out where I am going to stay in Paris and what I am going to do there. The new traveler approaches their trip as an experience first. Let's say somebody wants to go surfing or mountain climbing, or they want to see wild dolphins. They will go online with the experience first and discover where these experiences are available. This is an opportunity for countries and regions with wild nature. If you have a delicate wilderness that needs protecting, such as the arctic or a rainforest or clean coastlines, you may not want masses of travelers to come and interfere with the fragile environment. Nature photography is an industry that is taking off as people long for authentic nature experiences with wildlife. Digital alternatives such as live webcams and virtual reality experiences are ways to monetize nature with minimum interference.

If you've ever been to an investment promotion conference, online or offline, you may agree with me that those events are quite boring. There's a number of countries who all claim to have everything you need, without actually having done much research on who their customer is. The marketing message highlights available land properties, available workforce, possible tax cuts on energy and profits. They also often claim to be a logistical hub. Whether it's the link between Europe and Asia, the link between North America and Europe or the link between Asia and the South America, someone is claiming it. This is an old and tired story and if it has ever worked in the past, it certainly does not work now. If you genuinely feel that you are located in a real logistical hub you need to go deeper. You need to identify exactly who the offering is for and how they can benefit from your offering. Is it time saved, is it money saved, how and how much exactly? Being a logistical hub is a good story if you can frame it right. Covid-19 proved to the world how fragile our sourcing chains are. Ports in logistical hubs all over the world were (and still are) struggling with ships floating miles offshore, waiting for their turn to offload cargo. There were people in the ships suffering from Covid, people in the ports suffering from Covid, and people driving trucks and operating trains suffering from Covid. If you can truly claim that you have an offering that solves these massive global logistics problems you have gold on your hands.

KEY TAKEAWAYS:

1. There are ways to capitalize pure space.
2. Move up the value chain in R&D for natural resources.
3. It is possible to create industries around pure, wild nature.
4. If you market your areas as a logistical hub, make sure the offering is sustainable and consider each scenario.

PEOPLE AND INDUSTRIES: LABOR

Now let's go back to the resource called labor, your manpower, your labor force.

The top industries in the world in 2022[172] as measured by the number of employees are consumer electronics industry, commercial real estate, fast food restaurants, HR and recruitment, apparel manufacturing, hotels and resorts, mining, and tourism. The biggest industries in 2021 measured by market size are the financial services, construction, commercial real estate, eCommerce, Insurance, IT, the food industry, oil and gas, automobile manufacturing, and telecommunication.

But these two lists are like looking in the rearview mirror, as we know that automation will be replacing many of these existing jobs, and that technological advancements and the climate crisis will affect the market sizes of many on the second list.

To get a sense where labor will end up, we must start by examining where markets are going and how industries are shifting. For this latter, we can look at where investment capital is going. Cloud computing, biotechnology, data analytics, AI, real estate, green energy, pharmaceuticals, the Internet of Things (IoT), information security, and all types of sustainability-related industries are receiving the largest chunks of

172 Global Biggest Industries by Employment in 2022, IBIS World

private capital. Quantum technologies, advanced manufacturing including 3D printing, printed electronics, photonic chips or nanomaterials, and 5G edge computing are the rising fields of the future. Gartner has been issuing their annual emerging technology Hype Cycle since 1995. It is a great way to represent which technologies are rising, and where they are in their lifecycle. The 2022 Gartner Hype Cycle identifies 25 emerging technologies which they categorize in technologies that expand immersive experiences, accelerate artificial intelligence (AI) automation, and optimize technologist delivery.

Regardless of what your existing industry baseline looks like, these are trends that will affect your companies in the near-term and long-term future. In technologies in the first category that expand immersive experiences, you will see things like decentralized identity. This could be

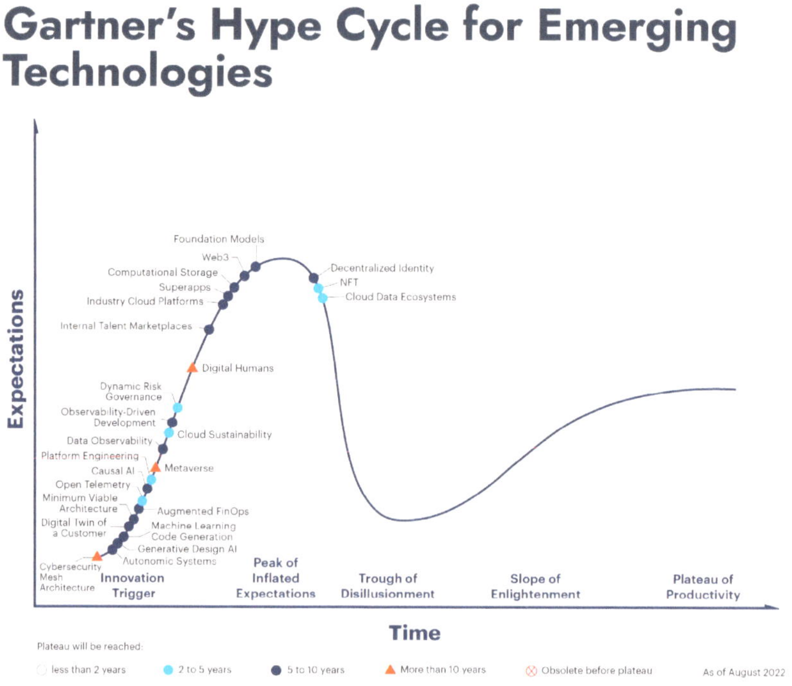

Source Gartner.

Gartner Hype Cycle for Emerging Technologies

a blockchain-based digital identity that allows people to safely operate online, holding digital wallets, non-fungible tokens (NFTs), and carrying out transactions safely. You're also going to see AI-driven "humans", digital humans that operate as a real human online representing personality, characteristics and a mindset of a human. Humans and digital humans operate in the metaverse as avatars and in Web3 as trusted identities.

Under the second category of accelerated AI automation, Gartner lists causal AI which can identify and utilize cause and effect relationships to predict relationships beyond correlation, and foundation models that embody a type of deep neural network architecture that computes a numerical representation of text. This will result in natural-sounding AI generated written and spoken language. Generative design AI is the use of AI, machine learning (ML) and natural language processing (NLP) to automatically generate design, content, and user flows. Machine learning code generation will enable AI to create code based on human's natural language description.

In the third category, optimized technologist delivery, Gartner describes a group of technologies that enable building digital businesses. These technologies provide feedback on your product, service or solution that enable people to build highly optimized, tailored, targeted products fast and in a sustainable manner. An example of such a baseline include cloud data ecosystems, that we already have in place but not in such sophistication. These technologies include generic sounding titles, but when implemented provide a powerful platform: Augmented FinOps that automate the financial governance of traditional DevOps, cloud sustainability, computational storage, cybersecurity mesh architecture (CSMA), data observability, dynamic risk governance (DRG), industry cloud platforms, minimum viable architecture (MVA), observability-driven development (ODD), OpenTelemetry, and Platform engineering.

These complicated sounding terms and acronyms will gradually change the way our current agriculture, food industry, hospitality, traffic and logistics, mining, manufacturing, or construction *workers will operate* in 2 to 10 years.

For someone to say they don't need to pay attention to technology trends is like being the proverbial frog in cold water, sitting and waiting for the water to heat up then boil them out of a job. I have been typing this book for about one year on Google Docs, and just the amount of AI driven assistance that Google has rolled out to improve the writing experience is mind boggling.

Climate crisis and technology will reform all of our current industries and thus the industry-specific skills the workforce will need to bring to the table. Some of these industries are more forward-looking in adapting than others. Coursera, a global online learning platform, provides two interesting studies on this. First, their 2021 study[173] on skills by industry studies 10 industries, across 26 skills in the Business, Technology, and Data Science domains. This study includes data points from the 82 million learners on the Coursera platform, and data from over 2,000 global businesses. The most future-proof industries that have invested in skills across business, technology, and data science are manufacturing, financial services, automotive, professional services, telecommunications, technology, energy and utilities, insurance, healthcare and consumer goods. Breaking this down into the three skill domains, we can see that the manufacturing industry has invested well in business and technology skills, and are rising in their investment level in data science skills. Finance has invested well in data science and technology skills, and are increasing in their business capabilities as well. The automotive industry is doing well overall in their skills across the three domains, whereas technology and telecommunications seem to be slipping on this chart.

The second Coursera study that I read is about talent. The Coursera Global Skills Report 2022,[174] which now uses data points from its 100 million online learners worldwide summarizes current events as the "great reskilling." This report splits global talent into four categories based on the overall skill level: Cutting-edge, Competitive, Emerging, and Lagging. Most of the regions fall into these four categories based on GDP, i.e. richer countries have better skill sets. But it is not entirely black and white. For Cutting Edge skill, the top quartile talent are mainly found in Europe and parts of Asia Pacific, where average GDP per capita is $33,975. There

173 Industry Skills Report, Coursera, 2021
174 Global Skills Report 2022, Coursera

Skills proficiency ranking 2021

The arrows indicate significant movement from the 2020 skills proficiency ranking

Rank	Business	Technology	Data Science
1	Manufacturing	Manufacturing ▲	Financial Services
2	Professional Services	Financial Services	Automotive
3	Automotive ▲	Technology ▼	Technology ▼
4	Financial Services ▲	Automotive ▲	Insurance ▲
5	Telecommunications ▼	Telecommunications	Manufacturing ▲
6	Energy and Utilities	Professional Services	Healthcare
7	Consumer Goods	Healthcare	Energy and Utilities
8	Healthcare	Insurance	Telecommunications ▼
9	Insurance	Energy and Utilities	Consumer Goods
10	Technology ▼	Consumer Goods ▼	Professional Services

Data Source Coursera 2022. Image credit Petra Soderling.

Skill sets valued in different industries

is still room for improvement by investing in skills that are required for innovation and next-generation technology. The second highest quartile, the Competitive skilled countries are the United States, parts of Southeast Europe, and parts of Asia Pacific. These regions have an average GDP per capita of $20,337, and their opportunity for growth is in the upskilling of workers to take advantage of new job opportunities. The Emerging category includes Latin America, Middle East and North Africa, and can improve by investing resource extraction revenue into diversified skills training. Lastly, the regions Lagging in skills are sub-Saharan Africa and the Caribbean. Lagging regions can improve by improving the capacity of workers to participate in the digital economy.

If we go back and look at the five countries in Section two—Brazil, Canada, Finland, Italy, and South Korea—they fall under two of these global skills performer buckets. Canada and Brazil are in the Emerging category, and Finland, Italy and South Korea in the Cutting-edge quartile.

Brazil is number 63 in the global ranking. Their skills in business are 16% of the top global talent ceiling, with technology skills at 73%, and data science skills at 34% (of the 100% maximum score). Brazil has been able to educate a pool of technology talent that far outperforms other skills in their country. The agriculture industry is suffering from global warming and economists fear this will have a big impact on Brazil's economy. Technological skills are the key to a resilient economy, in this regard the country has been doing wisely by investing in those skills. As the Brazilian economy faces headwinds from inflation, drought, supply-chain disruptions and the pandemic, technological innovation has the potential to spread prosperity. Coursera points out that climate technology requires data science, where Brazil is currently performing in 34% of the potential 100% ceiling. As the world's fourth-largest emitter of greenhouse gasses, and criticism about cutting down the Amazon rainforest in exchange for farmland, Brazil is well-positioned to lead the world in reducing its environmental impact. A good opportunity for Brazil is to invest in the intersection of agritech and climate tech—finding innovative solutions to address everything from resource extraction by using IoT sensors, satellite-based earth observation technologies, and livestock production with the support of biotechnologies.

There are two agritech developments that excite me. One is the intersection of agriculture and energy. Agrivoltaic farming—the addition of solar panels on top of farmed fields or planting crops under existing solar panels—is a way to help both activities. The solar panels bring shade to the plants reducing the need for irrigation, while also keeping the panels cool increasing their efficiency. Solar panels can also be installed over irrigation canals and reservoirs to decrease evaporation. The other area that is exciting is the intersection of agriculture and AI and is ripe with benefits. As farming equipment and satellite-based sensors collect more data, AI helps to create very precise fertilizing and irrigation plans that can change daily if not hourly. One of the biggest problems in agriculture, in addition to drought, is overwatering. Humans don't really know how much water plants need, and often use large-scale sprinklers that water too much. 20-40% of the water is overused in agriculture, and thus wasted. With detailed data, AI can calculate and release the exact amount needed to those plants who need it when they need it. In areas where water is used for generating power, saving water increases energy production. Also by not running irrigation pumps in vain, farmers can save energy.

In the case of South Korea, their global skills ranking across business, technology and data science is number 26. Business skills are 75%, and both technology and data science skills are 68% of the 100% maximum score. It is perhaps quite surprising that their score in business is higher than scores in technology and data science. Consumer electronics manufacturing being one of the big global industries in number of workers, market size, and in their ability to adopt new technologies, Korea has a bright future ahead if they stay on this path.

Other Asia Pacific countries that perform better than Korea are the region's number one and global number three country of Indonesia. Indonesia has invested heavily in worker training and education, and performs 100% in both technology and data science. Their business skills are only 30%, however. Singapore's scores are 90% in business, 95% in technology, and 82% in data science.

Finland, just like Korea, is in the Cutting edge skills category. Global number 13, Finland's skills in business are 65%, technology 91%, and data science 98% of the 100% maximum score. With electronics and machinery as Finland's top industries, its path is similar to South Korea's. They possess deep pockets of knowledge in future fields including quantum technologies, imaging and photonics, and applied AI.

Just like Brazil, Canada, Finland, Italy, and South Korea, your country or your state has something that you might not recognize. Call it your unique selling point if you will.

KEY TAKEAWAYS:

1. Emerging technologies will disrupt every industry.
2. There are big differences in global industries and how they are prepared for this.
3. Any country can get ahead by implementing focused skills development and management.

ACCESS TO CAPITAL: THE MONEY TO MAKE IT HAPPEN

And lastly, back to the input called money—capital.

Lending and borrowing creates creditors and debtors. It comes down to trust on both sides. Do I trust this person, company or country to pay me back if I lend them money? Do I trust in their capability to take my money and make more money?

Luckily, countries are not like individual people or companies in that they have extraordinary capabilities to manage leverage. As described in Section 1, some countries that hold reserve currencies can even print money. When it comes to countries, it's not a question of if they have access to capital, but how expensive that capital is. Governments can borrow money from private international banks, the International Monetary Fund IMF, the World Bank, the European Investment Bank, or from the markets by issuing government bonds. With private lenders, the interest rate and other terms are tied to the country's credit rating.

The countries with the best credit ratings at the moment—Australia, Canada, Denmark, Germany, Sweden, Switzerland, Norway and Hong Kong—all have something in common. They established a strategy that leaned on their historical strengths, their natural resources and their people, and they have been able to bring it together in a way that has paid off. Once again we see the same countries, those with the most democratic decision-making processes, the best education, the least corruption, and so on. These qualities have fed into a solid economic success which translates, among other attributes, to good credit rating.

The European Investment Bank (EIB) is owned by the EU Member States and is one of the largest lenders in the world. The EIB finances projects in both EU and non-EU countries that achieve the policy aims of the European Union. These policy aims include areas in climate, environment, small and medium sized enterprises (SMEs), development, cohesion, and infrastructure. In fact, The EIB is one of the biggest providers of green finance in the world. In 2007, the EIB became the first institution in the world to issue green bonds, and in 2019 it committed to stop funding fossil fuel projects by the end of 2021. The EIB's long-term loans are typically up to 50% of a project's overall cost, starting from €100 million to public sector entities.

International Monetary Fund (IMF) financing is meant to help countries tackle temporary crises, be they balance of payments problems, stabilization of their economies, or restoration of sustainable economic growth. The purpose of the loan is to provide temporary breathing room while the country is adjusting policies and restoring economic stability. IMF financing is often provided in response to natural disasters or pandemics. The IMF was established in 1944 in the aftermath of the Great Depression of the 1930s to ensure international economic cooperation in hard economic times. It is a member-based organization with 190 member countries. Each member country pays a capital subscription[175] when they become members. This is a quota that is based on that country's relative position in the world economy. Countries can then borrow from this pool of funds when they fall into financial difficulty. The IMF lists its assets as SDRs, Special Drawing Rights, instead of a central bank currency such as euro or dollar. SDRs are supplementary foreign exchange reserve assets defined and maintained by the International Monetary Fund. They represent a claim to currency held by IMF member countries

175 Where the IMF gets it money, International Monetary Fund

for which they may be exchanged. The current IMF reserve quota is SDR 477 billion which is equivalent to US$651 billion. The IMF also facilitates borrowing arrangements directly between member countries, with the total value of all of these arrangements at 973 billion SDRs, or more than US$1.3 trillion.

The IMF's toolkit for lending[176] is continuously refined to meet countries' changing needs. They are tailored to different types of balance of payment needs and to different membership circumstances. There is a General Resource Account (GRA) that is accessible to all members, and a Poverty Reduction and Growth Trust (PRGT) which is better tailored to the diversity and needs of low-income countries. These two resources are often used for improving and balancing economic instability. For both emerging and advanced market economies in immediate crises, the bulk of IMF assistance has been provided through Stand-By Arrangements (SBAs). The SBA addresses short-term payments problems. The Standby Credit Facility (SCF) serves as similar short-term relief for low-income countries.

For low-income countries facing protracted balance of payments problems, and in need of medium-term support, the Extended Fund Facility (EFF) and the corresponding Extended Credit Facility (ECF) are IMFs main tools. Their use has increased substantially since the global financial crisis in 2022. To help prevent or mitigate crises and boost market confidence during periods of heightened risks, members with already strong policies can use the Flexible Credit Line (FCL) or the Precautionary and Liquidity Line (PLL). Ukraine has used its quota, and these other instruments during 2022 represented emergency disbursement worth $4.9 billion. In October 2022, the IMF was preparing to carve out US$3-4 billion per month to support Ukraine's economy in 2023.

Other tools for low-income countries include the Rapid Financing Instrument (RFI) and the corresponding Rapid Credit Facility (RCF). These tools are designed to help with urgent balance of payments needs, including from commodity price shocks, natural disasters, and domestic fragilities.

176 IMF Lending, International Monetary Fund

While the IMF is the organization supporting economic stability in times of crisis, the World Bank is the go-to for development projects. Just like the IMF, the World Bank is made up of 189 member countries. These member countries are represented by a Board of Governors, which usually consist of countries' ministers of finance or ministers of development. The World Bank board holds an Annual Meeting of the Boards of Governors together with the International Monetary Fund. These two organizations ensure there are no gaps or overlaps in member countries' need for support.

The World Bank Group is made up of five independently organized units: The International Bank for Reconstruction and Development (IBRD), International Development Association (IDA), International Finance Corporation (IFC), Multilateral Investment Guarantee Agency (MIGA), and the International Centre for Settlement of Investment Disputes (ICSID). The first two, IBRD and the IDA, form what we know as the World Bank.

In fiscal year 2022, IBRD had over $30 billion in commitments to developing projects around the world. These commitments are carried out via instruments that support the different needs countries have in any given moment: Investment Project Financing for building physical and social infrastructures, Development Policy Financing that supports policy and institutional reform, and Program for Results Financing (and as the name suggests, is tied to a predefined outcome). 99% of the Development Policy Financing includes climate-related benefits, and 85% contribute to narrowing gender gaps.

Countries don't need to stick to these three instruments, however. They can combine elements of each and structure their own set of linked operations under one program. This is a new way of working for the World Bank which was created to rapidly mobilize US$18 billion as a response to the Covid-19 emergency. The country-specific programs can be vertical or horizontal where vertical is for support of just one country's long-term growth, and horizontal for multiple countries in a similar development need for a medium-term. The World Bank also has specific tools to ensure that small countries, those with less than 1.5 million people, are not left behind. In 2022, twenty-six small states received $2 billion in support.

Top borrowers from IBRD are India, Indonesia, Colombia, Morocco, and Argentina. We've seen Indonesia and Argentina perform well in the various categories we've discussed throughout the book, so these programs appear to be doing the job they are designed to do.

The second part of the World Bank, the International Development Association (IDA), is the world's largest multilateral source of concessional financing for the poorest countries. It provides financing in the form of development loans, grants, and guarantees to help these countries increase economic growth, reduce poverty, and improve living conditions for the poor. In 2022, IDA held almost $38 billion in commitments for 305 operations in agriculture, education, energy, financial sector, health, industry and trade, ICT, public administration, transportation, and water and waste management. The largest recipients of IDA services are Nigeria, Bangladesh, Democratic Republic of Congo, Ethiopia, and Kenya.

So where does the World Bank money come from exactly? IBRD finances its loans from its own equity and from money borrowed in the capital markets. To raise capital from the market, IBRD issues bonds. As they are rated AAA by Moody's and AAA by Standard & Poor's, investors view it as a safe investment for solid long-term return. IBRD can also borrow money against its assets which makes it a flexible organization. IDA is primarily funded by contributions from high- and middle-income partner countries, transfers from IBRD and other Bank Group institutions, borrowers' repayments of earlier IDA credits, and funding raised in the capital markets. IDA's credit rating is AAA.

If the World Bank can collect money by issuing bonds, so can individual countries. Global bond markets are actually larger than the collective stock markets, both in the monetary value of bonds outstanding and in dollar value of bonds traded daily. For investors, buying a bond is a low-yield but secure investment. In 2022, bonds from the US, UK, Italy, Germany, Australia, New Zealand, and Spain were yielding over 3% in secondary markets. If inflation were at 2%, your money would be safe and churning out modest but sustainable interest. In issuing a bond there are several things to consider[177]:

Maturity: This is the time that your bond is in the market until it expires and it makes its final payment.

177 What are government bonds and how do you trade them?, IG International

Principal: This is the amount the bond agrees to pay the bondholder at maturity.

Bond price: The issue price of a bond defines how much money your government receives for each bond. In theory, this should be the same as the principal. However, the financial markets have come up with secondary markets for bonds which adds a layer of complexity.

Coupon dates: Coupon dates are the dates on which the bond issuer is required to pay the coupon (the interest). It could be annually, semi-annually, quarterly or monthly.

Coupon rate: The coupon rate of a bond is the value of the bond's coupon payments expressed as a percentage of the bond's principal amount. For example, if your principal is $1000 and your coupon rate is 2%, the audience buying your bond receives $20 each coupon date.

Regardless of the current economic or political situation, there is a funding source for you. I described here international sources for large-scale projects, but your local and regional financial institutions are always available. For projects that benefit both your public economy as well as the private sector, public-private partnerships (PPP) are a good way to get commitment from all parties to invest and benefit from. The private companies can be local or international. Combining project-based public funding with Foreign Direct Investment efforts is a good win-win strategy for a city, state, or country.

KEY TAKEAWAYS:

1. Every tax-collecting region has access to leveraged capital.
2. Developing and underdeveloped countries have access to patient money.
3. Government bonds offer source of capital for the issuer and a safe investment for the bondholder.

HOW TO BRING IT TOGETHER

I moved to New Orleans Louisiana in the US in 2015. Being from Finland, and having only lived in New York City before that move, the culture shock was real. The south was exhilarating, fun, scary, inspiring, sad, and fascinating. Being a fresh transplant, I started to soak up the local history and culture. I visited museums, listened to music, made friends, ate food, drank cocktails, hung out on the porch, and read books. Books by, for, and about the south. About New Orleans. During the six years I spent time on the city's East and West banks, I gained knowledge and wisdom that I never would have received if I had stayed in the safe Caucasian, Protestant world that I grew up in. I'm grateful to New Orleans. I have my own version of the city, and while some proud locals may say it is a wrong one, it's mine.

During those years, I started a technology business there which raised some eyebrows, since New Orleans is best known for its music, art, and hospitality and not tech startups. Against conventional wisdom, I was able to hire and train many wonderful people who picked up the tech skills and performed their jobs with creativity and superb custom service. Sometimes it's good to bring in the unconventional wisdom of an outsider to see who you are.

There is a beautiful thought process of this in French philosopher Simone de Beauvoir's work where she reflects her own existentialism to the ideals of German philosopher Georg Wilhelm Friedrich Hegel,[178] the father of the dialectical scheme 'from thesis to antithesis to a synthesis'.[179] In de Beauvoir's Ethics of Ambiguity (1946), humans are constantly oscillating between facticity and transcendence. In accordance with her existentialism, a person is restricted by the boundaries of his existence (facticisism), yet he always aspires to overcome this situation and to create himself, to be free and transcendental. The human is a battleground of constant antithetical movements, movements between the thesis and the antithesis, which de Beauvoir calls *ambiguity*. Ambiguity is a characterization of human existence. We are ambiguous beings destined to throw ourselves into the future while simultaneously it is our very own existence that throws us back to our limited existence, our facticity. In a way, we are always destined to fail as pure and sustained transcendence. It is because of this fundamental failure that we realize that we cannot simply refer ourselves onto the world; instead, we must adhere to the common destiny of humans as an interrelated being. Hegel goes so far as to say that only because someone else recognizes me as a subject that I can be constituted as such. Outside the moment of recognition there is no self-consciousness. De Beauvoir takes to heart the Hegelian lesson and forms it into an ethics.

The challenge in front of you is enormous, but your job is not to resolve these issues for the whole world. Your job is to ensure your constituency is supporting the movement and among the first (if not the very first) to benefit.

Geopolitically, we are moving away from a globally-interconnected world which was built on factor-cost arbitrage and cooperative economic rules to a multipolar world with global connectedness coexisting with increased polarization. Globalization is a slogan from the past, but institutions like trade agreements and global financing remain. Your role is to be there defining how they work in this new era. The multipolar connectedness can also be called trusted connectivity, and it has been formed into a slogan "Ideology trumps economy" with regards to western companies being too dependent on sourcing Chinese components.

---

178 T. Malcolm Knox, Georg Wilhelm Friedrich Hegel, Encyclopædia Britannica

179 Douglas Burnham and George Papandreopoulos, Existentialism, The Internet Encyclopedia of Philosophy (IEP)

In technology, we are moving away from being connected on digital platforms, and being able to work and have fun on our laptops or phones. The digital experience is very soon to be a transversal, omnipotent invisible power that guides everything humans do on this planet. Whether we are building roads, ships, mining minerals, herding cattle, this is happening in a post-digital world with the use of AI, bioengineering, blockchain, quantum and edge computing.

Aging and declining populations in the west and in China and India present challenges to the economic development planner. An older population has a great demand for health services, which drives urbanization. The trend of urbanization of the aging clashes with the trend of young families wanting more room and a healthier environment, which lead to conflicts of interests within countries. While health issues can be addressed by use of new technologies, internal inequality will need to be addressed by the tools that democratic governments have.

We are at the end of an era with global access to fossil fuels that damage the planet, and quickly adopting affordable and feasible transition to low-carbon energy. The global infrastructure debt, including energy and transportation, that is now being recognized together with tectonic geopolitical shifts on energy, present two imperatives for governments to act immediately.

In the capital markets we've lived through an era of high debt and low inflation, allowing for economies to grow. Asian countries have enjoyed decades of extreme growth which cannot continue as disruptions settle to normalcy. The rising debt has been balanced off in our bookkeeping with physical assets built on earth minerals and fueled by fossil energy. As this model is coming to an end in the midst of Covid and geopolitical enabled sourcing problems, and billions of young Africans are entering the global market, we need to find a way to balance the bookkeeping with ourselves and with mother nature.

"But I am only one person. What can I do?" you may fret. What you can do, is to start being the change.

If we summarize what was learned in Section One and Section Two, we can conclude that the most successful countries in economic terms are the ones who pass ethical laws, create equality among citizens, drive an active fiscal and monetary policy, reach out internationally to make trade agreements, have a defined and structured immigration policy, ensure a living wage, invest in infrastructure, collaborate with the private sector on matters of national security, educate their citizens and encourage and fund startups. Success stories are born when these actions are married with local strengths, local environment, and the local psyche.

In this very last section, how to bring it all together, I want to round up the idea that innovation, economies, and industries being an interdependent trio that you can ignite and harness for an upward cycle.

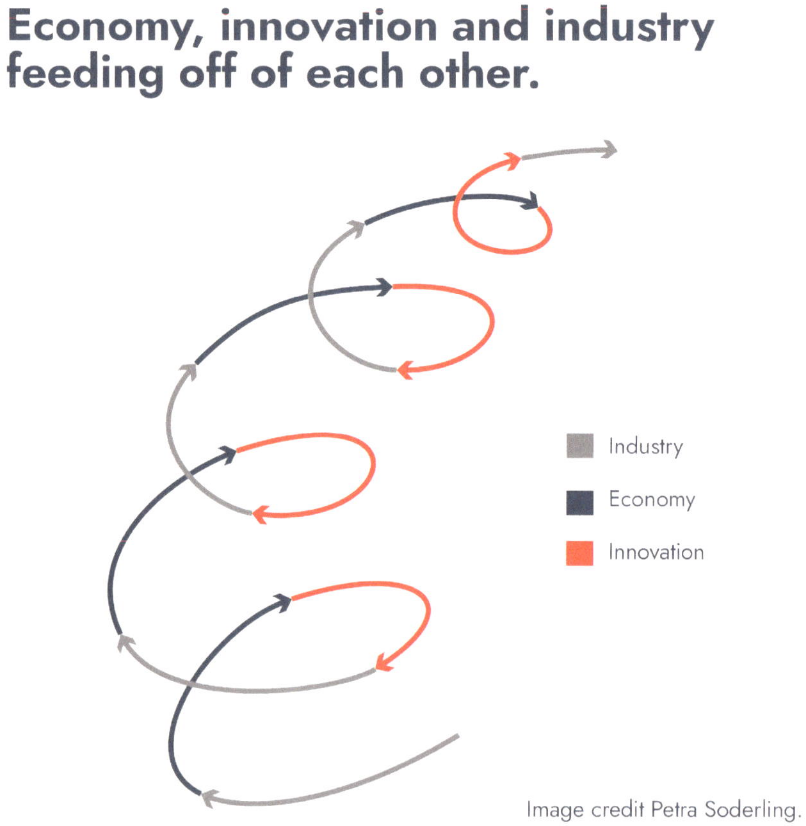

A virtuous upward cycle of innovation, industry growth, and economic growth

Economy is represented here by government. The government can use the tools and resources in their toolbox to create innovation. Innovation takes place everywhere, but for our purposes we argue this happens within publicly- or privately-funded universities, startups, and small- and medium-sized businesses. This innovation presents itself in the form of commercial products that enable companies to grow and entire new industries to be born. These successful industries contribute to their economy, making it stronger and wealthier, again able to invest into new innovation. The government is the key here. None of these other players have that "loaded" tool kit that can make this cycle happen.

This is why the next two chapters are about taxpayers. They are the ones who makes it all possible. You as an economic developer, governor, member of a parliament or congress, civil servant do not start from scratch. You are there because of taxpayers. They were here before you, and will be here after you. You have the power to take what they are offering you and turn it into something their children can enjoy. That's your job.

KEY TAKEAWAYS:

1. Zooming out and accepting the view of externals can help visualize path forward.
2. Geopolitical and environmental trends force their way into economic development planning.
3. The triangle of government, innovation, and business create an upward cycle.

TAX PAYERS: YOUR RISK INVESTORS

We pay taxes as individuals and as companies. Every time we use a service or buy something there are taxes involved. Somehow, however, we are being made to think that this tax is just on the sidelines of the economy and is not actually central to what the economy is all about.

When you turn on the TV to watch the news, you might see a politician talk about jobs, or an analyst talk about the stock market. Sales tax is something that is just on the receipt, not on your evening news. If taxes are in the news it is usually a debate about who should pay more taxes, the rich or the poor. You very rarely see an intelligent public conversation about taxes and how they relate to and influence what happens in the 'real' economy. Where is the tax money coming from, where is it going, what does it buy—and why is it needed?

There's a particular type of flow chart, the Sankey diagram, that shows the flow of tax money from left to right from the source of money to the use of money. The example I'm using here is from Minnesota. In Minnesota, as in most states and countries, the source of money is mostly from individual people's income tax and sales tax from their purchases. The terms used here are 'revenue' and 'expenditure'. For budgeting purposes these are perfectly good terms, but if we take a step back, shouldn't we

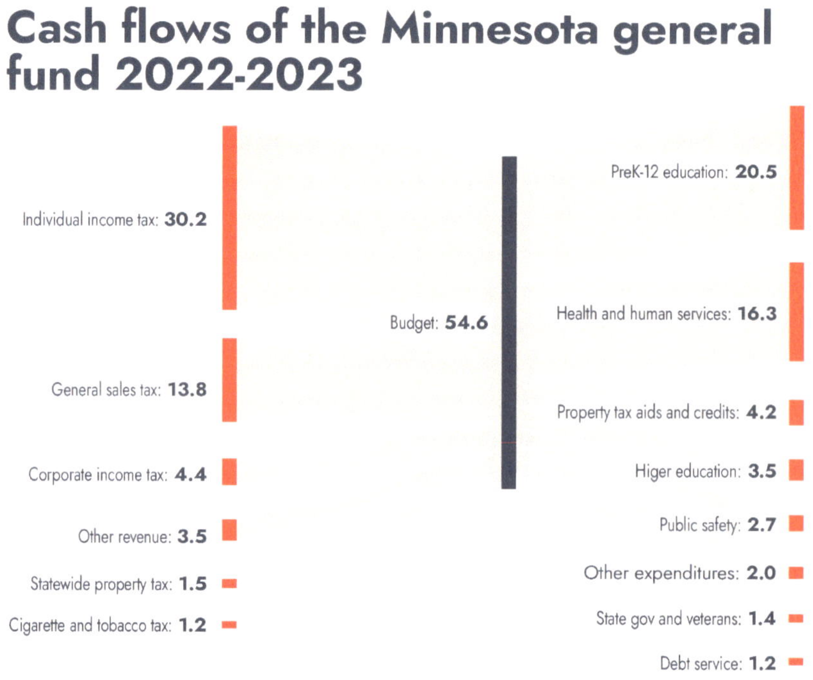

Example of a Sankey diagram, sources and uses of public budgets

be looking at "investment" instead? The taxpayers are investing in high quality education and healthcare. They are investing in public safety and infrastructure. All of these investments will bring a return in the form of healthy, educated citizens and a functioning environment where people can live and work comfortably—generating more tax dollars to be reinvested in a new generation.

Only a very corrupt system would be using money for something that would not yield prosperous results for the society or the economy. So why then is everyone talking about government spending instead of government investment?

There are multiple industries that got started by a government initiative. We've listed a few already: the space industry, the internet, the energy sector, media, telecommunications, and the defense industry. Come to think of it, only a very few original industries have been created by entrepreneur initiative. If we take the United States as an example, before the original settlers came from Europe, there was very little trade and no manufacturing. When the Europeans came they started first with farming and building, activities that would sustain them, yes, but that also required raw materials. This was the start of commerce in America but it was driven by individual needs and individual supplies. The first government investments to support this small-scale commercial economy were the postal system and the railroad, with a telegram service and ocean shipping coming next.

Some of you might think that this is the expected role of government, but is it really? What is the role of government? Is the role of a government to support the well-being of the people or is the role of the government to support the well-being of commerce? And are these things one and the same or two different things?

As we have established, the most successful capitalist democracies are the ones where the government collects the most taxes and invests the largest portion back into the economy. These are the countries with cutting edge tech companies and the best scientific R&D that looks into the fuzzy future and discovers potential new industries. A country with a very low tax rate can only dream of such success (unless of course you struck oil).

It is the natural next step to see the taxpayer as a risk inventor. The myth of 'government should not pick winners or losers' is busted.

If we think of the taxpayer being a risk investor as a first principle, it makes more sense[180]. The first principle is the basic assumption that cannot be deduced from any other proposition: It is the unmoved mover or the prime mover[181] that appears in mathematics, philosophy and in physics. First principle thinking is deconstructing a complex problem into fundamentals until the point when you can no longer break them up. Then, you start building from there. When you strategize innovation-based growth for your constituency, be a first principle thinker, and start with the taxpayer.

---

180 James Clear, First Principles: Elon Musk on the Power of Thinking for Yourself
181 Unmoved mover, Wikipedia

As a risk investor, you will want to create a strategy that focuses on your strengths, improves your weaknesses, and is diverse enough to ensure a sufficient return. In the private sector, a risk investor is often synonymous with a venture capitalist. A venture capitalist (VC) is someone who invests in high risk companies in exchange for equity, i.e. a stake in the company. There could be other types of risk investors, investors that don't take equity but rely on other types of return on their investment. For simplicity, as VC is the most familiar form of a risk investor, let's see how VC's do it.

Assuming the VC has raised their capital, they have a fund which they can invest, they need to find the investment target. For a VC this means generating a deal flow, or a deal funnel. According to a recent HBR survey[182] that interviewed almost 900 VCs, 30% of VC deals come from leads from their former colleagues or work acquaintances, and another 30% by VCs discovering entrepreneurs on their own. The rest, 20% are referrals by other VCs, 10% are cold email pitches by the entrepreneurs, and 8% are leads provided by their existing portfolio companies.

For each deal that ends up in investment, the VC looks at 101 opportunities on average. This is a staggering number, 100 rejected ideas for one accepted idea. Out of those 101, twenty-eight lead to a meeting with the entrepreneurs, ten of those will be reviewed in the VC partner meeting, 4.8 will receive due diligence, 1.7 leads to negotiating a term sheet, and finally one will actually sign the term sheet and receive funding. On average this will take 83 days, 118 hours work from the VC.

How is successful VC investment measured? The HBR study found that few VCs use standard financial-analysis techniques to assess deals. Most of the VC portfolio companies fail or yield very little, but some rare cases can bring a 100-fold return on its investment. Therefore most VCs just use this simple cash return as a measure of success. Business tools such as Net Present Value or Internal Return of Investment are tools for investors in a less risky environment. These types of tools are rarely used in risk capital, and if they are then after the fact. After an exit has been done, either as taking the company public with an IPO or by selling the company.

182 Steven N. Kaplan, and Ilya A. Strebulaev, How Venture Capitalists Make Decisions, Harvard Business Publishing, 2021

VCs also typically look at both the team and the idea. When pressed, ultimately the team is more important, as it is easier to tweak the idea or product with the right team than it is to replace a bad team to continue building a product someone else invented and is vested for. In the survey 95% of the surveyed VCs said that the team is an important factor, 74% said the business model was important, 68% highlighted the market, and 31% the industry.

So what do VCs ask in return for their money? That one entrepreneur out of the 101 that a VC reviews, will receive a Term Sheet that outlines all the asks that the VC has. In addition to the amount of equity (the share of the company), these usually include things like pro rata investment rights, liquidation preferences, and anti-dilution rights (which protect their potential economic upside) as well as on the vesting of the founders' equity, the company's valuation, and board control (which is often seen as the most important control mechanism). Pro rata rights, the right for the VCs to acquire an additional stake in a company, is often paramount because this allows for the VC to take a bigger share quickly if it looks like this case will be one of the winners. Other things you may see in the term sheet are an option pool to be given to employees or other partners, participation rights, investment amount, redemption rights, and dividends.

Now that we've learned how the VC operates, how does this compare with what the government does? Both governments and VCs invest in technologies and businesses with the expectation they will succeed. VCs invest money and their time as they advise their portfolio companies. Governments invest money, time, and enablers such as an educated and healthy workforce, flexible financial support for the company and their employees, help with cross-border exports and imports of parts and materials, and infrastructure.

Looking at dealflow, VCs both go out and seek investment targets, and receive referrals. National, regional, and local governments are more concerned about the enablers than direct dealflow, the bottom of the innovation pyramid if you will. However an innovation funnel is something that a successful country or a local economy should be concerned about. How your industries are growing, diversifying, and renewing themselves is not just a concern for the industries and companies, but for the community as a whole. Governments need smart industry strategies, i.e. a risk investment deal flow funnel. This deal flow funnel operates under the same laws of uncertainty as any private deal. The investor can strategize and optimize for the most favorable outcome but external risks always remain.

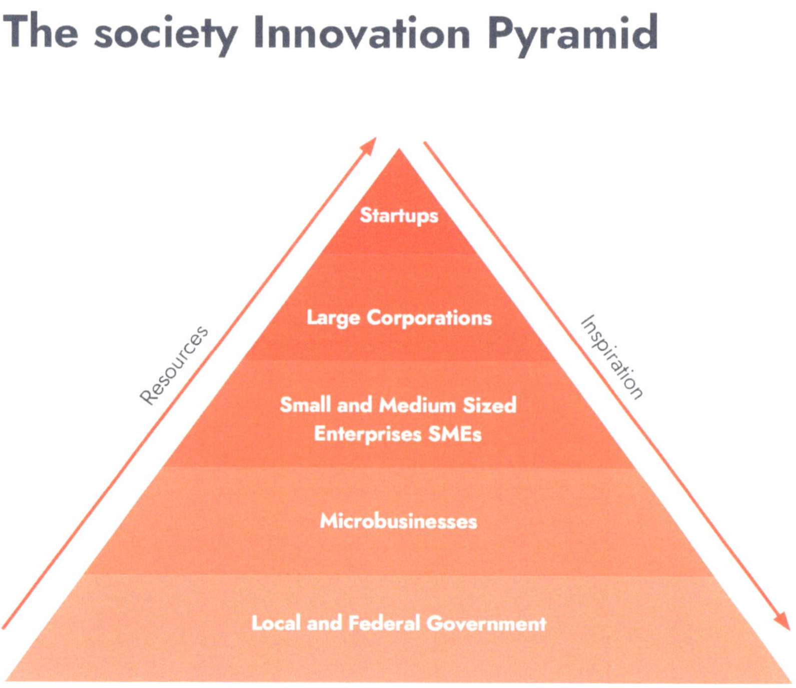

The Society Innovation Pyramid by Petra Soderling

Resources flow from public to private, innovation and inspiration from private to public

According to the HBR study a VC looks at 101 opportunities before investing in one. The conventional wisdom says that 1 in 10 invested portfolio companies is such a great success that it pays for the complete or partial failures of the other 9. This 2022 study[183] gave us these more concrete numbers:

183 Bobby Chernev, What Percentage of Startups Fail?, Review 42, 2022

- 90% of new startups fail.
- 75% of venture-backed startups fail.
- Under 50% of businesses make it to their fifth year.
- 33% of startups make it to the 10-year mark.
- Only 40% of startups actually turn a profit.
- 82% of businesses that fail do so because of cash flow problems.
- The highest failure rate occurs in the information industry (63%).

The failure rate of government risk investments is harder to calculate as the investment (the input) as well as the return (the output) is more diverse, more long term, more integrated and interwoven into the society and its different actors. The part of VC investing in both team and idea is similar, where a government is investing in its people and it is investing in science. The return, the reward is the same though. It can be measured in financial value.

My argument is, even though these two risk investment types are not identical, they share similarities that should be celebrated, not criticized. Those startups that failed ended up creating something, their contribution didn't just vanish. Maybe the product lives on in another form, in another company, maybe the product died but the team learned valuable lessons and are now successfully driving their ideas forward in a new, more successful company. During their lifetimes, the failed startups paid salaries and taxes and helped families move their lives forward. In a similar manner, when a government invests in a field of research that isn't immediately generating GDP, it is still a valuable part of the economy. A field of science may take decades to mature and when it does, the rewards are hundreds-thousands-or millions-fold. Artificial Intelligence started receiving DARPA and other US government funding in the 1950s; we are seeing mainstream commercialization today. Cybersecurity research funding took off at ARPANET in the 1970s; all major digital players invest in cybersecurity tools and tactics. Early quantum computing that relies on even earlier quantum mechanics research as well as nuclear reactions—these and other long-term investments in their early days were risk investments. Risk is in the sense that one in a ten or one in a hundred would succeed so well financially that it would far offset the "failure" of some other fields that were not commercial successes. Although as you can see I put failure in quotes, science has value far beyond commercial success; thus, I'm only using this term here in the context of investment.

So as you see, government investments over time pay off in the short- and long-term in ways that we cannot even understand in the moment of allocating those investments. Public investment is at least as productive as private, and several strands of the research seem to indicate that it is substantially more productive.[184] I'm drawing this as the society innovation pyramid multiplying into infinity over time. There is no way of knowing who is building or where and when success is built based on public investments we are making here today.

The Society Innovation Pyramid over time

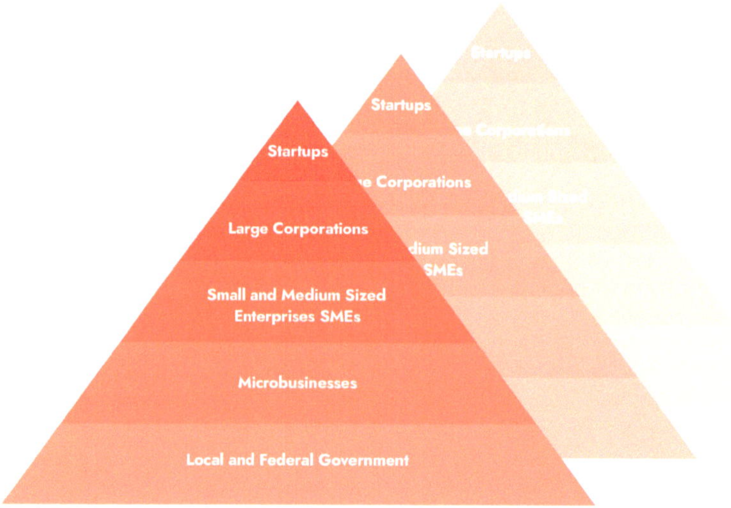

The Society Innovation Pyramid by Petra Soderling

Resources flow from public to private, innovation and inspiration from private to public. Over time this creates sustainable innovation based economic movement.

---

184 Josh Bivens, Public investment, The Economic Policy Institute, April 18, 2012

KEY TAKEAWAYS:

1. Taxes keep societies running.
2. VC investments and tax-paid investments share synergies and interdependencies.
3. Leveraging these synergies and interdependencies will create a stronger society innovation pyramid.

PROTECTING OUR INVESTMENTS

When the World Bank assessed what Brazil had done right with their agriculture R&D investments, one of the spotlights was on mission orientation and IPR (intellectual property rights) policy. Technology transfer is equally as important as technology creation. "Pursuing an open innovation system and IPR policy in the agricultural sector facilitated technology transfer, diffusion of new cultivars, and the filing of international patents." The World Bank also commented on what Brazil was right not to do. By discouraging just any curiosity-driven research, their government research lab was able to produce inventions solving real problems in the Brazilian agriculture industry that ended up in commercial use. The World Bank was, in other words, complimenting the Brazilian government for setting a strategy, goals, and a field of practice for publicly-funded researchers. Embrapa, as we learned, creates intellectual property that it owns and licenses out to private companies, and creates shared or joint IP with partners. The Brazilian government liked this idea and ensured the country has strong IPR in agriculture research. This attracted international companies to come and do agritech in Brazil, adding to the Foreign Direct Investment pool.

Publicly-funded intellectual property needs protection against misuse and exploitation, just as a privately-funded IP does. As the investors, taxpayers have the right to ensure their property is being used for the good of all of the people—not just a few within their country or only people outside of their country.

Earlier, I said that whether the government should ask for ownership in exchange for its support is an ongoing conversation. There are good arguments for and against it, and different countries in different stages of their development have taken different approaches. Some argue that by creating government-owned spinoff startups from government-owned universities or research labs, a developing or underdeveloped country has a chance to quickly boost the commercialization of research and enhance the use of homegrown innovative technologies by their domestic private companies. This argument also supports the commercialization of technologies that are of national security interest. For example, it helps a government if their local private industries are protected by domestically-invented and -sourced cybersecurity or AI solutions. On the other hand, a good argument is made for governments not to retain equity in startups, as this could be a red flag for any international private investors. A government owner could potentially overrule any Shareholder Agreement or Bylaws, or even if not that severe, a private investor might fear that the government owner has a different exit strategy than the risk investor has. A government owner might also scare off customers in the fear of being sucked into a public procurement process instead of a business-to-business transaction.

The USA is the best country today for protecting private companies' intellectual property rights (IPR). The International Property Rights Index 2022 [185] placed the USA as #1 in IPR and Finland as #1 in physical property rights (PPR). According to the index, the assignment of intellectual property rights does not grant exclusive possession (which physical property rights do), but the benefits of its economic exploitation, promoting the generation of economic incentives towards research and innovation, as well as stimulating the open exposure of ideas, encouraging indirect effects of creativity benefit the society as a whole. A strong property rights regime, on the other hand, promotes people's confidence in its effectiveness to protect private property rights. It offers an integrated, effective and efficient

185 International Property Rights Index 2022, Property Rights Alliance

USA and Finland leading the Intellectual Property Rights Index

Image credit Petra Soderling.

The Intellectual Property Rights Index lists leaders in Intellectual Property Rights (IPR) and Physical Property rights (PPR)

system for registering your property, and it allows access to the required credit to become an owner or to convert that property into capital.

The United States is historically a global anomaly in its willingness to protect individual inventors. There is even an intellectual property clause in the Constitution, passed unanimously, with the intent to "promote the progress of science and useful arts, by securing for limited times to authors and inventors the exclusive right to their respective writings and discoveries." The young republic understood the importance of technical innovation, and much debate was held around the ways of ensuring best gains for the inventor and for the country. Congress recommended that the government offer grants to innovators in exchange for monopoly licensing for the government of their technology, and the government would also attach conditions to these grants such as maximum prices that the IP holder could charge, minimum quantities of said product to be produced, and conditions under which employees would be producing these products. These policies were tested over the 18th and 19th centuries with inventions and public and private enterprises in maritime and shipping, aviation, and the early telecommunications industry that started with the telegram. There are many fascinating stories of these early experiences in "The Battle over Patents - History and Politics of Innovation" by Haber & Lamoureaux (2021) if you are interested in more reading.

Eventually the US settled for a non-ambiguous market-based patent system where the inventor retains rights for their invention and is free to capitalize it in the market without government intervention. Any inventor could patent any invention as long as it was not fraudulently achieved, US Congress would not interfere what types of inventions would be made and published, and if they would generate commercial interest or not. The British and French model, an administered system, retained the right for the government to grant monopolies to inventors, guiding what types of inventions and businesses would emerge, and who would be running them. The American model was not free from inventions becoming monopolies however, as a patent or a group of patents and other intellectual rights properties could easily result in a de-facto monopoly, as happened with Western Union and AT&T, for example. The US would fight monopoly through antitrust legislation.

What happened over a hundred years ago should not be taken as a guidance for action today. There certainly are innovation-based technology monopolies in the USA and in the entrepreneur-led startup culture in Europe today.

In addition to national governments, public universities are authorities in assigning and securing intellectual property. In the US, there is a history of universities wanting to retain ownership of the invention, especially as licensing out this IP is a major source of income to many institutions. This ownership has been secured with individual contracts between the university and its professors and researchers. The Bayh-Dole Act, and a subsequent Supreme Court case 'Stanford University v. Roche Molecular Systems, Inc.' from 2011, reaffirmed that in the United States the invention belongs to the inventor. Companies and universities can offer employment contracts where the employee gives up this right, and they do. If you are an inventor in the USA and you are looking to secure rights to your invention, I would recommend going at it alone with private investors. Uncle Sam will make sure you're protected.

The same practice is prevalent in most western countries, except the Nordic countries, which have held a tradition called professor's privilege, where university researchers are able to retain full rights to their inventions that they are working on while employed at the university. Finland, Denmark, and Norway have all changed this model as part of harmonizing for pan-European practices, going from the professor's privilege to a more US-based model where the university owns all intellectual property rights to the inventions. Sweden still keeps the model that favors the professor.

Finland changed from professor's privilege to university ownership in 2007. A study[186] about inventors and patenting in Finland between 1995 and 2010 suggest that patenting by university-employed individuals dropped 46% from the announcement of the change, and by 29 percent after the change in 2007. Researchers Ejermo and Toivanen (2018) summarize that "Our and others' results call into question whether the European reform of the professor's privilege were good innovation policy." Other studies[187] found that allowing universities to claim ownership of inventions made by their researchers and professors led to a "decline in the technological importance and the value of patents owned and managed by universities." [188] Also, patented research results in these countries have been moving from universities to private companies, and the importance of companies as innovation drivers has been increasing at the expense of academia.

Retaining IP rights or taking equity in spinoff startups aren't the only ways to protect your investments. Protecting our public investments can also happen through:

- Laws: Anyone betraying your country by selling secrets to foreign adversaries will face consequences.
- Standards: By inserting your preferential IP into global essential standards you ensure everyone wanting to build products or

186 Olof Ejermoab and Hannes Toivanen, University invention and the abolishment of the professor's privilege in Finland, Research Policy, 2018

187 Francesco Lissoni, Peter Lotz, Jens Schovsbo and Adele Treccani, Academic patenting and the professor's privilege: Evidence on Denmark from the KEINS database, Science and Public Policy, October 01, 2009

188 Catalina Martínez and Valerio Sterzi, The impact of the abolishment of the professor's privilege on European university-owned patents, Industry and Innovation, January 09, 2020

services on top of this technology will have to use your IP and pay you for it.

- Trade policy: You can dictate that bad actors cannot operate in your country, your domestic companies cannot depend on supplies from an adversary, and citizens of your country are not allowed to work for adversary companies in adversary countries.
- Export controls: Ensure critical technologies are not exported to countries you don't trust.
- Sourcing channels: Ensure your domestic companies are not dependent on sourcing from adversary countries thus risking sharing critical information.
- Intelligence (CIA, FBI): Your national security agents are following events with adverse powers to track any misuse of your protected technology or IP.

I'm a big supporter of going with the standardization route. This is a non-restrictive, positive way of ensuring your government and your local companies get the compensation you are entitled to, with a need to invest less in spying and policing your competition and adversaries. Having others depend on key technical components will also give you other leverage, such as a good position in the global sourcing channels, and overall attractiveness as a destination for talent looking to leverage the good, solid innovation pyramid that you've built.

KEY TAKEAWAYS:

1. Strong protection of intellectual property created in your country accelerates innovation.
2. IPR governance affects Foreign Direct Investment.
3. Countries have a diverse set of tools to protect domestic intellectual property.

YOUR STRATEGY FOR LONG-TERM PROSPERITY

We need government strategies for critical technologies. The strategy should be focused around 1) removing obstacles from private companies to grow, 2) enabling completely new industries to be born, and 3) addressing existential threats to citizens and humanity. Scaling that to your constituency, I'm slightly modifying the Society Innovation Pyramid to better represent the work ahead of you. This government building is constructed from building blocks for your long-term economic success, and it rests first and foremost on your natural resources. You cannot change the size, shape, location, or nature of your land.

The base is a solid block of national, regional, and local government work that includes budgets, vision statements, and thousands of working hours that produce the enablers for people, companies, and academic institutions to freely do what they do best. Here you must ensure you have the needed education provided on all necessary levels, that you have a healthcare system that will allow for people to live and work carefree. You need taxation, infrastructure, and legal and jurisdictional branches that operate in a way that benefits the society.

Right above the government, I'm placing the vast number of micro-businesses that keep our economies rolling. This group of earners and

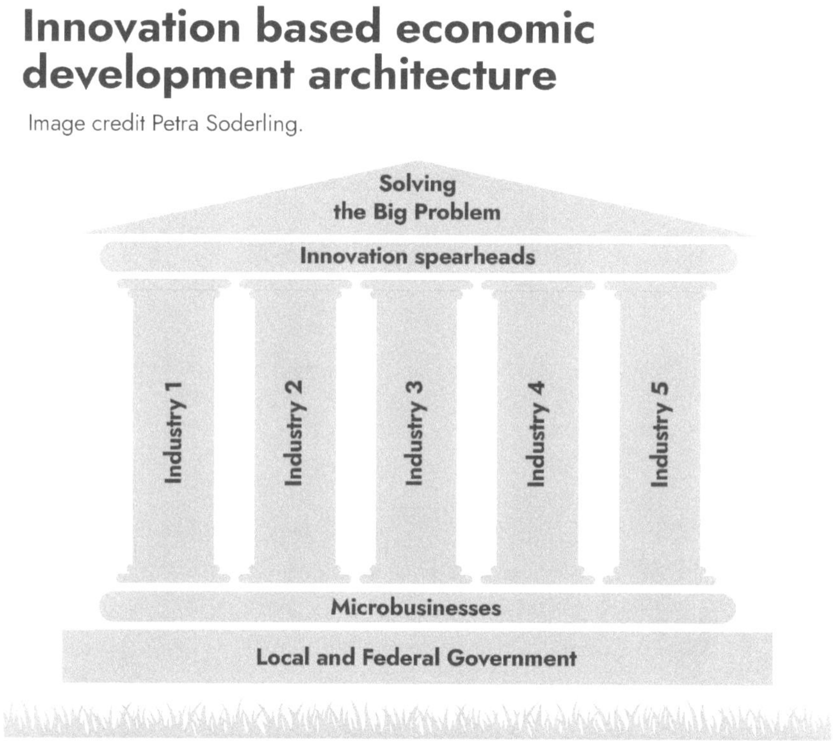

Innovation based economic development architecture

providers are a critical, but often overlooked, group. Think of the nannies and childcare workers we discussed in the immigration chapters, think of plumbers, landscapers, hairdressers, bookkeeping companies, florists, and cleaners. If it wasn't for these services the society would collapse. This floor in our building needs attention as much as the ceiling does.

The pillars that rest on the foundation and the floor holding up the ceiling are your main industries. These are often key export industries that bring in foreign currency, and keep up the trade. These industries offer jobs, pay taxes, and each in their own way contribute to our societies and economies. These pillars need to be able to operate trusting that their needs will be met, and nothing sudden or dramatic will disrupt their operation. The government needs to ensure these industries have the legal framework, international trade status, employees, and the infrastructure they need to be able to make long term plans.

The ceiling of our building consists of the innovators, disrupters, startups, scientists, and students who create forward-looking inspirational concepts, products, solutions, trends, companies and fields. It is this group of people who look at the roof—the future direction the world is going and what the future holds. Even though the innovators' gaze is upward and onward, they are standing on the shoulders of government, micro-businesses, and traditional businesses. Often we mistake their job to be fixing the problems in the foundation, asking for the innovators to mend an issue in a traditional industry, or in the society, but this is someone avoiding responsibility. As an example, if a social media company has created a situation where they are a threat to democracy, the lawmakers can certainly ask for the social media company to fix the problems inside their companies but it is the lawmaker who ultimately is responsible for democracy. In the same way, it is not the responsibility of a cleantech startup to clean up the pollution that a legacy fossil fuel company has created. They certainly can offer solutions and products that can help solve the issue, but ultimately the responsibility lies in the lower layer.

The top pediment, the triangle sitting on top of our innovation ceiling and industry pillars, is the bright future that we all aspire to create. This is how we weatherproof your economy to ensure your pillars and foundation will not get cracks and mold, and continue to support the system long into the future.

As they say, a good teacher doesn't fill the pot but lights the fire. I'm going to leave these last pages for you to fill out. After reading this book, and maybe some of the recommended resource reading, take a cup of tea and a pen, relax on the sofa, and start to strategize. Fill in ideas in the book, a separate workbook, or just in your mind for now. I hope these sparks will result in an economic development plan that is both realistic and down to earth, and ambitious and forward leaning all at once. I trust you.

Innovation based economic development architecture – government base

Image credit Petra Soderling.

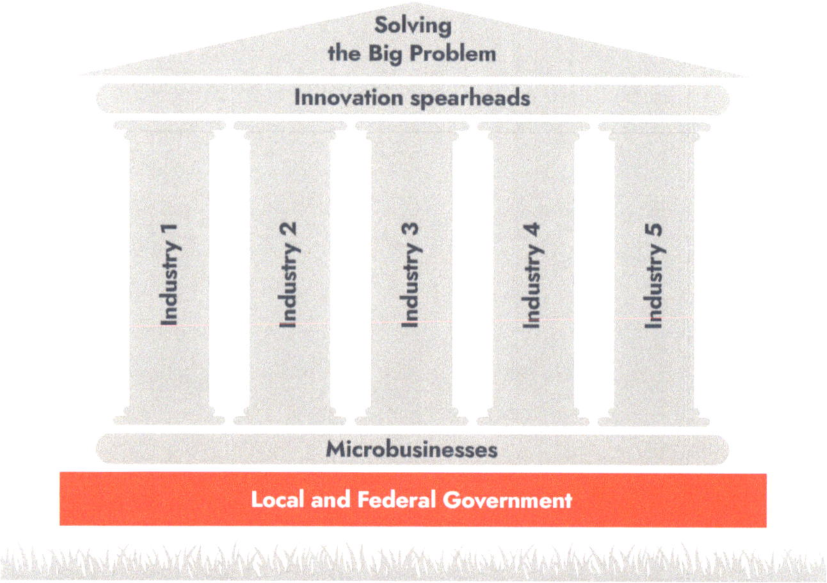

Innovation based economic development architecture – government base

As you identify and understand your natural resources, geographical position, your people and where you stand financially, what are the strengths within your team that you can utilize? Is there anything you need to outsource? How do the people in your community view your organization? Where are your strengths and weaknesses? Does your government need strengthening in some areas?

Innovation based economic development architecture – microbusinesses

Image credit Petra Soderling.

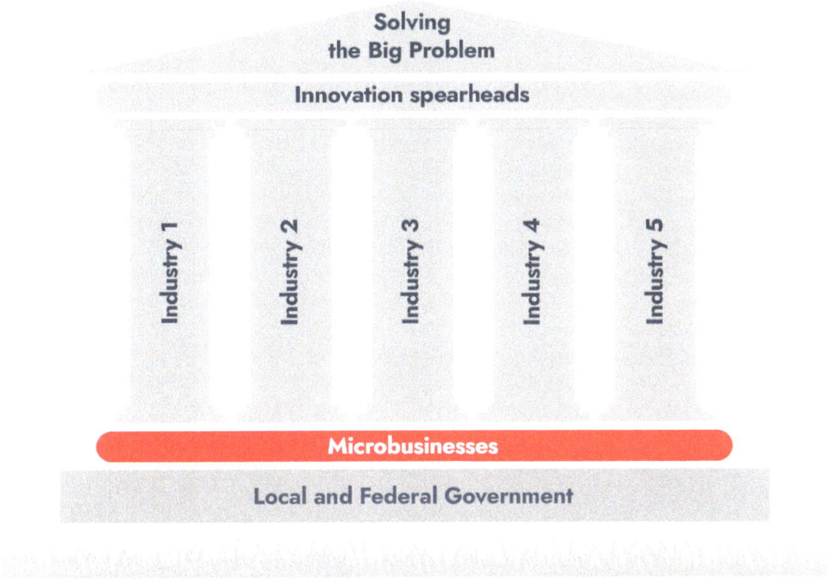

Innovation based economic development architecture – the microbusiness layer

Do small businesses have the freedom to innovate and thrive? Are you supporting them with wages, daycare, healthcare, or home-related support systems? Are the blue collar entrepreneurs and service industry microbusinesses doing what they love of their own free will, or are they forced to a job they don't enjoy because of financial distress?

Innovation based economic development architecture – industry vertical #1

Image credit Petra Soderling.

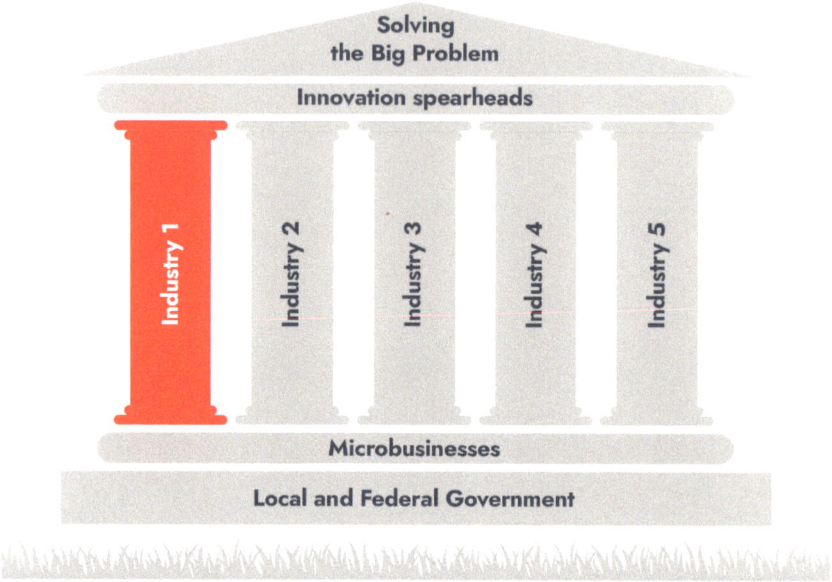

Innovation based economic development architecture – industry vertical #1

What is the biggest industry in your region, the one that employs most people and brings in a good chunk of the tax revenue. Maybe it's a prominent energy cluster, or one big vehicle manufacturer. Are you supporting them to innovate and renew? Are you helping them realize that even though things look good for them now, there is a disruption around the corner and you can help them address it head on?

Innovation based economic development architecture – industry vertical #2

Image credit Petra Soderling.

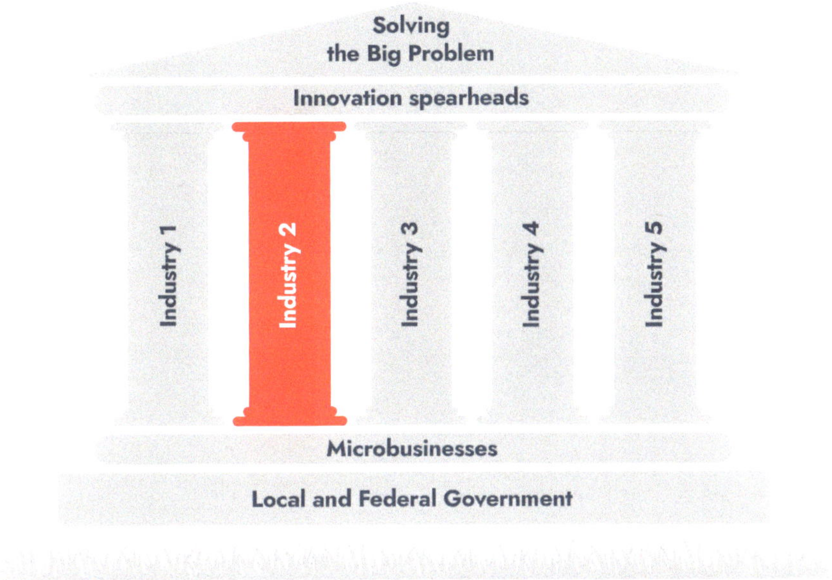

Innovation based economic development architecture – industry vertical #2

What is the second most important industry that you foster? Possibly a female-dominated service industry that is burdened by low wages and lack of vision for the future? Can you help with family related support, offering better healthcare, childcare, or low cost public transportation for this industry to thrive?

Innovation based economic development architecture – industry vertical #3

Image credit Petra Soderling.

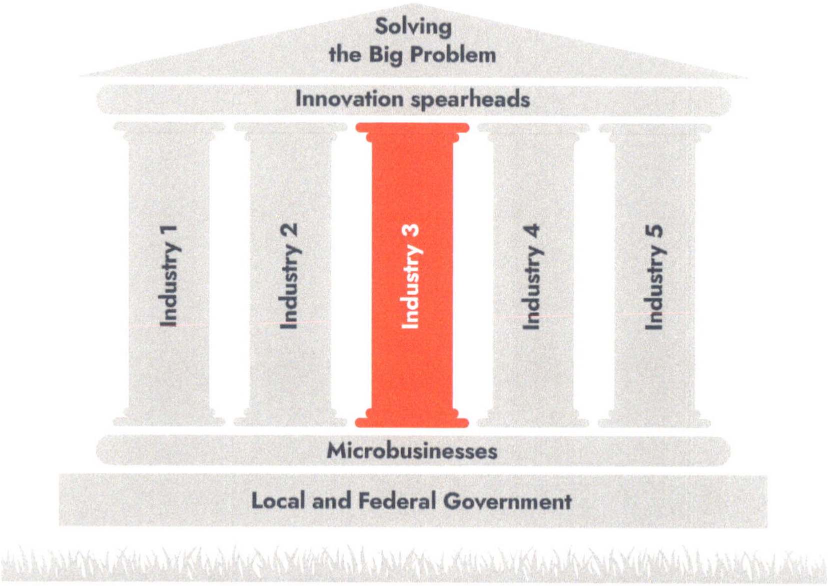

Innovation based economic development architecture – industry vertical #3

Who is your industry number three? Is it retail? What type of problems are they facing and how can you help? Digital platforms, global procurement channels, and the price of energy are certainly a concern to them.

Innovation based economic development architecture – industry vertical #4

Image credit Petra Soderling.

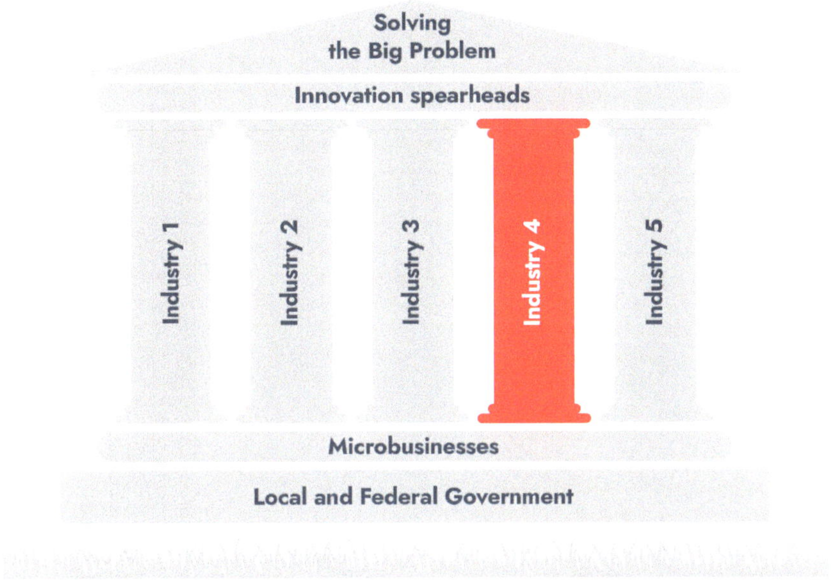

Innovation based economic development architecture – industry vertical #4

Which of your local or national industries do you rank as the fourth on your list to ensure will succeed? It could be ICT. Just because they create and deliver technology doesn't mean they are prepared for the future, as we read in the chapter on people and industries. You may be able to help with legislation, trade agreements, or by tailoring community college classes.

Innovation based economic development architecture – industry vertical #5

Image credit Petra Soderling.

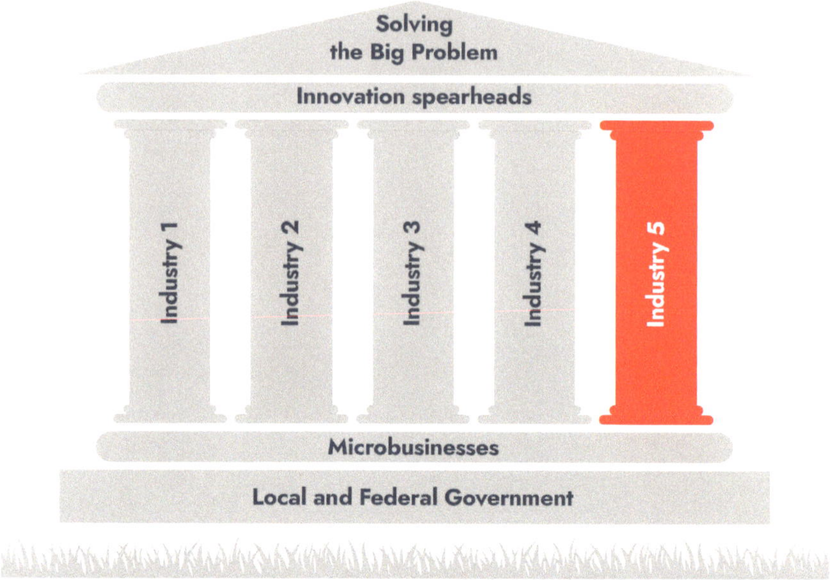

Innovation based economic development architecture – industry vertical #5

I'm limiting the number of industries here to five, but you can keep going. The fifth industry on your list may be one that hasn't been important in terms of economic development efforts but nevertheless cannot be left behind. Is it hospitality, healthcare, tourism, entertainment, or the creative industries? It is always dangerous for decision-makers to think that any industry will survive challenges because the people doing it love what they are doing and "would do it for free." The fact is no one can live for free, so these industries need special attention. Arts and culture are the food to our souls, and a soulless economy never survived in the history of humankind.

Innovation based economic development architecture – Innovation spear needs

Image credit Petra Soderling.

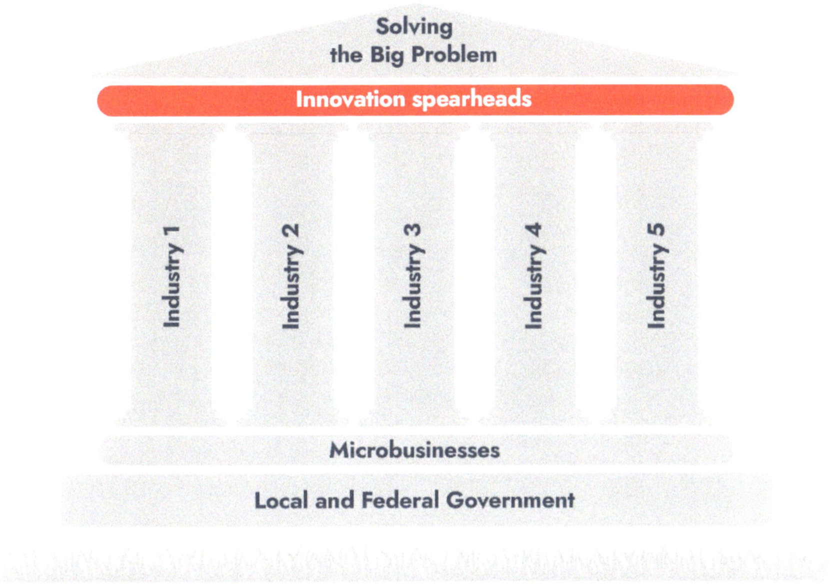

Innovation based economic development architecture – Innovation spearheads

The most innovative teams in your community are the startups, the students, and academia's researchers. What can you do to make sure this precious group of people stays with you, and stay innovative? They need resources, freedoms, and a trust that you will be able to carry them forward.

Innovation based economic development architecture – Solve the Big Problem

Image credit Petra Soderling.

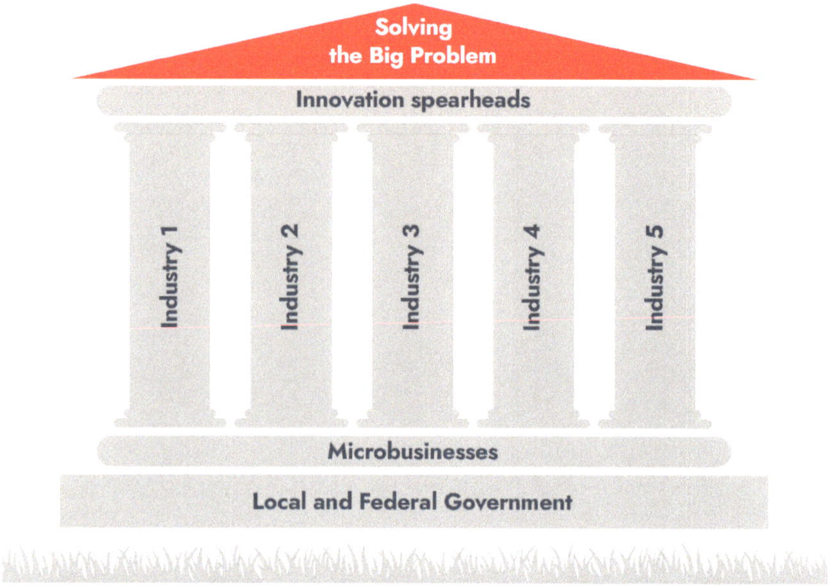

Innovation based economic development architecture – Solve the big problem

Finally, your biggest job as the face of the future of economic development is to ensure your citizens, students, workers, entrepreneurs, investors and shareholders see that you have a vision on how to carry this forward in the new era. We are at the end of an era that grew shareholder value and GDP by exploiting resources. What are you going to do about that? How can you pivot? What kind of sacrifices are you making to abandon the old in order to focus on where the future success lies? What are you not going to do moving forward? Where are you taking your voters and your taxpayers? How will they and their grandchildren's grandchildren benefit from your strategies?

As your colleague, I'm as puzzled as you are. There are so many questions and no certainty as to what the right answers will be. I'm looking at this from the history of my professional and private roles in Finland,

London, Silicon Valley, New Orleans, New York, Colorado, Serbia, and Occitanie France. Each region has its unique characteristics, developmental phases, and outlooks. Your strategy for the long-term prosperity of your community is in your hands.

CONCLUSION

One of the fundamental changes we need to make in economic development is to start planning for where the puck is going instead of where it is sitting now. We know that climate crisis is changing our natural environment; we know that AI is changing our online behavior; we know that a pandemic can bring the entire world to a standstill; we know that there are smart kids in developing countries who can do the same digital job as your citizens in your developed country for a fraction of the price. It's no longer just about good-paying jobs. Optimizing for jobs is optimizing for the present, which would be fine if we did not have competition. But we do.

Economic development is not one of the fields that college graduates ever flock to. We are the boring, invisible people behind press releases. During my years in private industry, I would tune out if someone started marketing their region for me. I'm ashamed to admit that I did not feel that they had anything to offer to me. What would someone who wants to create jobs in Virginia have to offer to me as a business development professional creating new products and solutions in the latest technologies. I was not looking for customers from Virginia. I was not looking for employees in Virginia. I wasn't looking for any particular technology that only Virginia could offer. Economic Development to me at that time felt like an industry that was exploiting the capabilities and findings that we, the technology industry, were creating. How wrong I was.

I started working in the Foreign Direct Investment field in 2018, creating investments from North American ICT companies to Finland, and the breadth and depth of this work opened my eyes. None of the large international companies establishing greenfield R&D operations or manufacturing facilities randomly decide to set them up in Malaysia or Germany or Columbia. Each decision has a thorough web of documents, promotion materials, plans, negotiations and policy decisions backing them up. Economic Development, Investment Promotion, and Foreign Direct Investment are serious commercial actors that are seriously undervalued.

The world is changing. 20th century rules do not apply in the 21st century—not in business, not in industry, not in economics, not in society, and not in public policy. We have to create new ways of thinking, new rules, and new strategies for the 21st century. It's high time for Economic Development to be raised to the standard to which it belongs. Ensuring that countries, states, cities, counties, communities have a voice in shaping how their lives, and making sure that businesses, workplaces and futures are being defined are critical activities for the survival of this planet. We live in a time of grassroots movements. Young people are actively promoting their view of their reality, racial inequality, gender-based discrimination, destruction of the planet, mistreatment of animals, pollution of the oceans and air. People care about these issues and they are no longer willing to compromise the well-being of humans and the well-being of nature for economic gains.

We who work to develop our regions have to invite this reality into our decision-making. Using the old ways of planning or measuring success are not going to cut it. In this book, we've discussed macro-level tools—laws, interest rates, trade deals, and other political instruments. These are all necessary tools for macro-level management but we also need other newer tools. We need new tools that anticipate the new future.

The new future is not nation-states hosting large monopolistic companies creating technological and legal silos to promote their business. The future is local economies providing for the needs of the local population. These local populations reside in different environments of the world. They could be in the arctic or the tropics, in the deserts or even on water. What is important to understand is the economic needs of these communities and populations. As we plan for economic development, we can no longer look at large land masses, we can no longer rely on large governmental plans for areas the size of Russia, China, India, or the United States. Economic needs should henceforth be split into and serve smaller groups or areas, while still supporting the global agenda. As our home planet is facing hardship in the form of a climate crisis, different regions suffer different consequences while facing possibly new opportunities. These opportunities are local or regional, so they cannot be planned on a national or federal level. We are going to need new types of economic communities that thrive based on local or regional resources.

Think about very dry conditions. There are plants which thrive in such conditions and produce extremely nutrient edible plants. I think of ancient vine species or tea plants. The more these plants suffer from a lack of nutrients and water, the more antioxidants they produce. At the other extreme, there are regions which suffer from excess water. These regions suffer from algae and fungi, insects and so on that may sound at first like nuisances, but can actually be cultivated into food, medicine, or new biomaterials.

I've had two ongoing thoughts during this writing process. The first is that there's always one single person behind every large movement. The second is that we have to be working from positivity, from a belief that solutions exist and that we can find them.

Therefore, I'm asking you to be the hero you want to see. If not you, who? If not now, when? You are planting seeds that will carry your community forward for years and decades to come. So you need to ask, "which seeds?" We don't have a crystal ball, so we need to rely on our best knowledge, understanding, intuition and empathy to make the right choices for the future. We won't be here to see all of the ripple effects but we can take comfort in the thoughts that these ripples are going to be doing something good.

I want to leave you with a poem by the first Poet Laureate of the United States, Robert Penn Warren (by permission of the Heirs of Robert Penn Warren). This is the final poem from his sequence of poems called *Audubon: A vision.*

TELL ME A STORY

A)

Long ago, in Kentucky, I, a boy, stood
By a dirt road, in first dark, and heard
The great geese hoot northward.

I could not see them, there being no moon
And the stars sparse. I heard them.

I did not know what was happening to my heart.

It was the season before the elderberry blooms,
Therefore they were going north.

The sound was passing northward.

B)

Tell me a story.
In this century, and moment, of mania,
Tell me a story.

Make it a story of great distances, and starlight.

The name of the story will be Time,
But you must not pronounce its name.

Tell me a story of deep delight.

ACKNOWLEDGEMENTS

Thanking these people doesn't seem like it could be enough, but … Thank You!

My Gratitude goes to:

Ismo Rantala, Vivian Norris, Danika Hannon, Tamma Ford, Marko Markovic, Gabrielle Gerbaud, Sophie Alcorn, Robert Hays, Mitsuko Klein, Tanya Ramond, Meeri Haataja, Jon Radoff, Robert Beletic, Olli Pekonen, Brian Lenahan, Rao Konidena, Hanna Hyttinen, Outi Keski-Aijo, Laura Thomas, Veera Semi, Yoonchae Cheong, Charlie Schick, Ceriweden Koski, Mart Aro, Matt Versaggi, Arfan Chaudhry, Luca Rossettini, the Heirs of Robert Penn Warren—and so many others who supported me in many wonderful and mysterious ways!

GLOSSARY

3G	Third generation cellular network technology
3GPP	3rd Generation Partnership Project
5G	Fifth generation cellular network technology
6G	Sixth generation cellular network technology
AFRL	Air Force Research Lab
AI/ML	Artificial Intelligence/Machine Learning
ARPANET	The Advanced Research Projects Agency Network
B2B	Business to Business
BBC	British Broadcasting Corporation
BoF	Birds of Feather
BRIC	Brazil, Russia, India, and China
BRICS	Brazil, Russia, India, and China, and South Africa
CAD	Canadian dollar
CEO	Chief Executive Officer
CERN	European Organization for Nuclear Research
CIA	Central Intelligence Agency
DARPA	Defense Advanced Research Projects Agency
DoD	Department of Defense
ECB	European Central Bank
EFTA	European Free Trade Area
ELDO	European Launcher Development Organisation
ESA	The European Space Agency
ESRO	European Space Research Organisation
ETLA	Elinkeinoelämän tutkimuslaitos, Economic Research
EUR	Euro
FAIA	First AI Accelerator

FBI	Federal Bureau of Investigation
FCAI	Finnish Center for Artificial Intelligence
FDI	Foreign Direct Investment
Fed	Federal Reserve of the United States
G20	The Group of Twenty
G7	The Group of Seven
GA	General Assembly of the United Nations
GATT	General Agreement on Tariffs and Trade
GDP	Gross Domestic Product
GEM	Global Entrepreneurship Monitor
GPS	Global Positioning System
HBR	Harvard Business Review
HF	High Frequency radio
HPC	High Performance Computer
ICT	Information and Communications Technology
IEEE	Institute of Electrical and Electronics Engineers
IIJA	Infrastructure Investment and Jobs Act
IMF	International Monetary Fund
IP	Intellectual Property
IPR	Intellectual Property
ISED	Innovation, Science and Economic Development Canada
ITU	International Telecommunication Union
K-12	Kindergarten through 12th grade, elementary and secondary school
LAN/WAN	Local Area Network/wide area network
LNG	Liquified Natural Gas
MBA	Master of Business Administration
MINUSCA	The United Nations Multidimensional Integrated Stabilization Mission in the Central African Republic
MINUSMA	The United Nations Multidimensional Integrated Stabilization Mission in Mali (MINUSMA), (MINUSCA), and
NAFTA	The North American Free Trade Agreement

NASA	The National Aeronautics and Space Administration
NQS	National Quantum Strategy
OECD	Organization for Economic Cooperation and Development
PPR	Physical Property Rights
PISA	Program for International Student Assessment
PPP	Paycheck Protection Program or Public Private Partnership
R&D	Research and Development
RCEP	Regional Comprehensive Economic Partnership
RDT&E	Research, development, test, and evaluation
RRR	Reserve ratio requirement of a central bank
SBA	Small Business Administration
SIDF	Saudi Industrial Development Fund (SIDF)
SME	Small and Medium Sized company
TTIP	Transatlantic Trade and Investment Partnership
UAV	Unmanned Aerial Vehicle
UHF	Ultra-High Frequency radio
UK	United Kingdom
UNESCO	United Nations
UNESCO	The United Nations Educational, Scientific and Cultural Organization
UNSOS	The United Nations Special Operation in Somalia
USD	US dollar
USMCA	The U.S. – Mexico – Canada Agreement
VAR	Vector Autoregression
VAT	Value Added Tax
VC	Venture Capital
VHF	Very High Frequency radio
VR	Virtual Reality
VTT	Valtion Teknillinen Tutkimuslaitos - the Technical Research Centre of Finland
WIPO	World Intellectual Property Organization
WTO	World Trade Organization

INDEX

D

E

F

G

R

S

T

U

V

W

X

Y

Z

RESOURCES

21 Lessons for the 21st Century, Yuval Noah Harari, Penguin Random House, 2019

23 Things They Don't Tell You About Capitalism, Ha-Joon Chang, Bloomsbury Press, 2010

A Promised Land, Barack Obama, Crown, 2020

Chaos theory discloses triggers and drivers of plankton dynamics in stable environment, IrenaV.Telesh, Hendrik Schubert, Klaus D. Joehnk, Reinhard Heerkloss, Rhena Schumann, Martin Feike, Arne Schoor & Sergei O. Skarlato, Scientific Reports, Nature, 2019

Chaos theory explained: A deep dive into an unpredictable universe, Paul Sutter, Space.com, March 18, 2022

Ethics of Ambiguity, Simone de Beauvoir, Open Road Media, Reissue edition, May 8, 2018

Formation and Evolution of the Knowledge Regime and the Development Process in Korea. Aymes, J. F. L. Korean studies 2014

Full-Spectrum Thinking - How to Escape Boxes in a Post-Categorial Future, Bob Johansen, Berrett-Koehler Publishers, 2020

Growth, from Microorganisms to Megacities, Vaclav Smil, The MIT press, 2019

Higher education and economic development in the OECD: policy lessons for other countries and regions. Journal of higher education policy and management. Pinheiro, R. & Pillay, P. 2016

How Will You Measure Your Life?, Clayton M. Christensen, James Allworth, Karen Dillon, Harper Business, 2012

Humankind, a Hopeful History, Rutger Bregman, Little, Brown, and Company, 2020

Industry and Technology Policies in Korea, OECD Publishing, 2014

Liquid gold the story of liquid crystal displays and the creation of an industry. Castellano, J. A. World Scientific Publishing Company, 2005

Mastering the Dynamics of Innovation, James M. Utterback, Harvard Business School Press, 1996

National Quantum Strategy Consultations: What We Heard Report, ISED, the Innovation, Science and Economic Development Canada, ©Her Majesty the Queen in Right of Canada, as represented by the Minister of Industry, 2022.

OECD Reviews of Innovation Policy: Korea 2009, OECD Publishing, Paris, 2009

OLED display fundamentals and applications, Tsujimura, T. (2017) Second edition. Hoboken, New Jersey: John Wiley & Sons Incorporated.

On the cusp of a new era? By Chris Bradley, Jeongmin Seong, Sven Smit, and Jonathan Woetzel, McKinsey&Company, October 20, 2022 |

Originals - How Non-Conformists Move the World, Adam Grant, Penguin Books, 2016

Principles for Dealing with the Changing World Order - Why Nations Succeed and Fail, Ray Dalio, Avid Reader Press, 2021

Prisoners of Geography - Ten Maps That Explain Everything About the World, Tim Marshall, Scribner, 2016

Public Investment: The Next 'New Thing' for Powering Economic Growth. Briefing Paper #338. Josh Bivens, Economic Policy Institute, April 2012

Quantum Canada, Ben Sussman et al 2019 Quantum Sci. Technol. 4 020503, IOP Publishing Ltd, 2019

Relative share of public, private and international expenditure on educational institutions, by final source of funds (2018), 16 Sept 2021, OECD

Review of international political economy: RIPE

Sapiens, A Brief History of Humankind, Yuval Noah Harari, Harper Perennial, 2015

Shadow Courts - the Tribunals that Rule Global Trade, Haley Sweetlands Edwards, Columbia Global Reports, 2016

Spectrum Wars - The Rise of 5G and Beyond, Jennifer A. Manner, Artech House, 2022

The 'MacGyvered' Weapons in Ukraine's Arsenal, Helene Cooper, Eric Schmitt, August 28, 2022, New York Times

The Battle over Patents - History and Politics of Innovation, Stephen H. Haber and Naomi R. Lamoreaux, Oxford University Press, 2021

The Entrepreneurial State - Debunking Public vs. Private Sector Myths, Mariana Mazzucato, Public Affairs, 2018

The role of the state as an inter-scalar mediator in globalizing liquid crystal display industry development in South Korea. Lee, Y.-S. et al. 2014

The Voice of the Poet, Adrienne Bond, Georgia Humanities Council, 1989

Thinking Fast and Slow, Daniel Kahneman, Penguin Books, 2012

Utopia for Realists - How We Can Build the Ideal World, Rutger Bregman, Back Bay Books, 2017

Why the Age of American Progress Ended, Derek Thompson, The Atlantic January/February 2023 issue

World Order, Henry Kissinger, Penguin Press, 2014

www.ingramcontent.com/pod-product-compliance
Ingram Content Group UK Ltd.
Pitfield, Milton Keynes, MK11 3LW, UK
UKHW061545051026
15625UKWH00026B/695/J
* 9 7 9 8 9 8 8 1 1 2 4 2 6 *